From Street-smart to Web-wise®

In a world where tiny fingers are as familiar with touchscreens as they are with crayons, ensuring our children's safety online has never been more crucial. *From Street-smart to Web-wise®: A Cyber Safety Training Program Built for Teachers and Designed for Children* isn't just another book – it's a passionate call to action for teachers and a roadmap to navigate the digital landscape safely, with confidence and care.

Written by authors who are recognized experts in their respective fields, this accessible manual is a timely resource for educators. Dive into engaging content that illuminates the importance of cyber safety, not only in our classrooms but extending into the global community.

Each chapter is filled with practical examples, stimulating discussion points, and ready-to-use lesson plans tailored for students in kindergarten through second grade. Regardless of your technology skill level, this book will provide you with the guidance and the tools you need to make student cyber-safety awareness practical, fun, and impactful.

As parents partner with educators to create cyber-secure spaces, this book stands as a framework of commitment to that partnership. It's a testament to taking proactive steps in equipping our young learners with the awareness and skills they need to tread the digital world securely.

By choosing *From Street-smart to Web-wise®: A Cyber Safety Training Program Built for Teachers and Designed for Children*, you position yourself at the forefront of educational guardianship, championing a future where our children can explore, learn, and grow online without fear. Join us on this journey to empower the next generation—one click at a time!

From Street-smart to Web-wise®

A Cyber Safety Training Program Built for Teachers and Designed for Children (Book 1)

Al Marcella, Brian Moore, and Madeline Parisi

CRC Press is an imprint of the
Taylor & Francis Group, an **informa** business

Designed cover image: © Shutterstock

First edition published 2025
by CRC Press
2385 NW Executive Center Drive, Suite 320, Boca Raton FL 33431

and by CRC Press
4 Park Square, Milton Park, Abingdon, Oxon, OX14 4RN

CRC Press is an imprint of Taylor & Francis Group, LLC

© 2025 Al Marcella, Brian Moore, and Madeline Parisi

Reasonable efforts have been made to publish reliable data and information, but the author and publisher cannot assume responsibility for the validity of all materials or the consequences of their use. The authors and publishers have attempted to trace the copyright holders of all material reproduced in this publication and apologize to copyright holders if permission to publish in this form has not been obtained. If any copyright material has not been acknowledged please write and let us know so we may rectify in any future reprint.

Except as permitted under U.S. Copyright Law, no part of this book may be reprinted, reproduced, transmitted, or utilized in any form by any electronic, mechanical, or other means, now known or hereafter invented, including photocopying, microfilming, and recording, or in any information storage or retrieval system, without written permission from the publishers.

For permission to photocopy or use material electronically from this work, access www.copyright.com or contact the Copyright Clearance Center, Inc. (CCC), 222 Rosewood Drive, Danvers, MA 01923, 978-750-8400. For works that are not available on CCC please contact mpkbookspermissions@tandf.co.uk

Trademark notice: Product or corporate names may be trademarks or registered trademarks and are used only for identification and explanation without intent to infringe.

ISBN: 978-1-032-73173-5 (hbk)
ISBN: 978-1-032-73784-3 (pbk)
ISBN: 978-1-003-46592-8 (ebk)

DOI: 10.1201/9781003465928

Typeset in Caslon
by codeMantra

To our families and friends,

for their encouragement, support, and love.

and

To the dedicated educators globally, who nurture,

guide, support, and ignite a passion for knowledge

and learning in children of every educational background.

Contents

List of Figures

List of Tables

Foreword

Today's students are exposed to online hazards at increasingly younger ages. The ability to navigate this digital landscape safely and responsibly has never been more important, which makes the timing for this program ideal.

As a former elementary school teacher and a current university professor with a background in education/curriculum development, I am sure that leaders who use this valuable resource will help our students develop the essential skills and awareness necessary to safely cope in our digital age.

The authors have a deep understanding of not only how children learn but what leaders need to help them do so. The necessary behaviors to protect children online are explained in an age-appropriate fashion and in a manner that will resonate with young learners.

The lesson plans are creative, thorough, and easy-to-use without relying on scare tactics or being overly prescriptive. This structure still allows leaders the flexibility to make the program their own. As a result, students will become more effective problem solvers and thoughtful decision makers while increasing their zest for learning.

Technology tends to move more quickly than our ability to keep up and protect users from unintended consequences. This underscores why a program based on cybersecurity best practices is so needed at this time. The program helps leaders navigate our rapidly evolving digital landscape while ensuring our students safely access the positive aspects of technology.

Anyone tasked with the responsibility of ensuring our student's well-being and online safety will find this resource indispensable.

John F. Boyce, Head – People Development, AMSOIL, Inc.

Preface

The authors developed this text for teachers and student leaders as the reader and user, and not the student who is the ultimate beneficiary of the material.

This teaching manual is designed to help make an educator's job and knowledge transfer easier by providing context to young learners and by providing relevant lesson plans for each chapter.

The approach is structured to emphasize the following:

- The creation of content in lesson plan format.
- The emphasis on skills training and associated assessments.
- The development of critical thinking skills regarding cyber hygiene which begins at the earliest ages.
- The design and development of instructional materials that can be delivered regardless of the instructor's or facilitator's experience level.

In a school classroom environment, it's often best to use inclusive and neutral language when referring to children and educators, as this promotes a respectful and supportive atmosphere.

The reader will notice throughout this book that instead of using terms such as "girl, boy, he, she, boys, or girls" or using gender-specific terms, the authors will refer to children when discussing pedagogy as "students," "learners," or "young learners."

These learners may be involved in any academic learning program, within any positive learning environment serving students.

When referring to teachers, the authors recognize that this descriptor is intentionally broad. The authors use the term teacher or educator to collectively include instructress or instructor, lecturer, tutor, facilitator, mentor, counselor, educationalist, and trainer – i.e. a person who regularly works with children.

Examples Are Not Endorsements

This document contains references to materials that are provided for the reader's convenience. The inclusion of these references is not intended to reflect their importance, nor is it intended to endorse any views expressed, or products or services offered by third-party providers. These reference materials may contain the views and recommendations of various subject matter experts as well as hypertext links, contact addresses, and websites to information created and maintained by other public and private organizations.

The opinions expressed in any of these third-party materials included as references or examples do not necessarily reflect the positions or policies of the authors.

The authors do not control or guarantee the accuracy, relevance, timeliness, update, or completeness of any outside, third-party information included in this publication. These references should not be construed as an endorsement by the authors or by the publisher. The reader should validate and substantiate any information directly from third-party providers before authorizing the acquisition, implementation, or use of any product or service referenced in this book.

Acknowledgments

The authors wish to thank the many contributors who have provided input to this program as it transformed from a concept with various delivery methods to this resource for teachers and those persons who regularly work with children, we thank you.

The authors wish to recognize and thank the following for their contributions and for so graciously responding to requests for further information.

John F. Boyce, Head-People Development, AMSOIL Inc. Duluth, Minnesota and Adjunct Faculty, Northwestern University, Evanston, Illinois, for providing insight into book development at the earliest stages of transferring the concept to a teacher's manual.

Christine Burke, Elementary School Principal, St. Louis, Missouri, for providing feedback to CRC Press which provided insight to the authors.

The authors also thank the anonymous reviewers identified by the publisher, CRC Press, who initially reviewed our proposed book series and provided positive feedback and comments on the proposed content when the authors initially brought the idea for this book to CRC Press.

Authors

Dr. Al Marcella, Ph.D., CISA, CISM, President of Business Automation Consultants (BAC) LLC, is an internationally recognized public speaker, researcher, IT consultant, and workshop and seminar leader with 46 years of experience in IT audit, risk management, IT security, and assessing internal controls, having authored numerous articles and 30 books on various IT, audit, and security related subjects. Dr. Marcella's clients include organizations in financial services, IT, banking, petrol-chemical, transportation, services industry, public utilities, telecommunications, and departments of government and nonprofits. Dr. Marcella is also a tenured, full-time professor at Webster University, teaching at the Walker School of Business.

Research conducted by Dr. Marcella on unmanned aircraft systems, cyber extortion, workplace violence, personal privacy, electronic stored information, privacy risk, cyber forensics, disaster and incident management planning, the Internet of Things, ethics, and astrophotography has been published in the *ISACA Journal*, *Disaster Recovery Journal*, *Journal of Forensic & Investigative Accounting*, *EDPACS*, *ISSA Journal*, *Continuity Insights*, *Internal Auditor Magazine* and the Astronomical League's *Reflector Magazine*. Dr. Marcella's fourth book on cyber forensics, *Cyber Forensics: Examining Emerging and*

Hybrid Technologies, a collaborative effort written with a team of six co-authors, was published by CRC Press in 2021.

Dr. Marcella holds a B.S. degree in Management, a B.S. degree in Information Technology Management, an M.B.A. with a concentration in Finance, and a Ph.D. in Management/Information Technology Management. Dr. Marcella is a Certified Information Systems Auditor (CISA) and a Certified Information Security Manager (CISM), and holds an ISACA Cybersecurity Certificate.

Dr. Marcella is the 2016 recipient of the Information Systems Security Association's Security Professional of the Year award and recipient of the Institute of Internal Auditors Leon R. Radde Educator of the Year 2000 award and has been recognized by the Institute of Internal Auditors as a Distinguished Adjunct Faculty Member. Dr. Marcella has taught IT audit seminar courses for the Institute of Internal Auditors (IIA) and the Information Systems Audit and Control Association (ISACA).

Brian Moore is a passionate Certified K-8 General Education Teacher with over 20 years of experience developing and implementing diverse curriculums covering a wide range of subjects. Highly skilled at motivating students through positive encouragement and reinforcement of concepts via interactive classroom instruction and observation, Moore is successful in helping students develop strong literacy, numeracy, social and learning skills.

Moore has valuable experience in classroom administration, professional development and project planning in one of Phoenix Arizona's largest Unified School Districts. Prior to the classroom, Moore was the site director for the Scottsdale/Paradise Valley (AZ) YMCA's Before and After School Program.

Equipped with extensive background in versatile education environments, a student-centric instructor, academic facilitator and motivational coach, Moore is well-versed in classroom and online technologies.

Moore received both his Bachelor of Arts and Master of Education: Educational Leadership from Northern Arizona University–Flagstaff, Arizona, and holds Certifications: Standard Professional Elementary Education, K-8 and Pre-K-12 Principal, and the Endorsement: Structured English Instruction, K-12.

Madeline Parisi, M.Ad.Ed., Principal, founded Madeline Parisi & Associates LLC (MPA) in 2013, an international organization providing business training content and training material, after a distinguished career working with professional associations. Together with a pool of subject matter experts, MPA provides business training materials, in-house and virtual training, white-label writing services, and professional certification training and certification exam question development services.

Parisi is a recognized adult education content developer with a 30-plus year career in business training and professional development serving finance, legal, and IT audit, IT security, and risk management professionals.

Parisi holds a Bachelor of Arts degree in Criminology from the University of Illinois, Chicago and a Master of Adult Education from National Louis University, Wheeling, Illinois. Parisi also holds certificates in Volunteer Management from Harper College, Palatine, Illinois and additional certificates in Organizational Development, Lean Six Sigma, and Project Management. Parisi is the author of several columns and articles published in various trade publications and along with Al Marcella is the co-author of the white paper "Assessing, Managing and Mitigating Workplace Violence: Active Shooter Threat."

Partnering with Al Marcella, The Training Resource Center, LLC, is a jointly operated entity providing training and consulting services, specializing in Environment, Social, and Governance (ESG).

1

Character Education

Introduction

The core mission of character education is to guide students to make thoughtful choices and act with integrity. Lessons emphasize universal values such as honesty, respect, responsibility, fairness, and compassion. The goal is to equip students with an ethical framework they can apply throughout their lives.

Quality character education is not an add-on but is woven into the entire learning experience. From modeling good behavior to discussing moral dilemmas, teachers can integrate ethics into all subjects. Schools may also create a culture that reinforces kindness, hard work, and community service.

Character Education

What Is Character?

Character can be defined in various forms, depending on personal contexts. Some examples include:

- Understanding, caring about, and acting upon core ethical values.
- The set of characteristics that motivate and enable one to function as a moral agent, do one's best work, effectively collaborate in the common space to promote the common good, and inquire about and pursue knowledge and truth.
- A set of personal virtues that produce specific moral emotions, inform motivation, and guide conduct.
- The traits and moral or ethical qualities distinctive to an individual.[1]

DOI: 10.1201/9781003465928-1

How Do We Define Character Education?

Character education refers to the deliberate effort to foster positive qualities and values in individuals to help them develop good character traits. It goes beyond academic instruction and focuses on instilling virtues, ethical principles, and social skills that contribute to personal and social well-being. Character education aims to shape individuals into responsible, respectful, and compassionate members of society.

Key components of character education often include:

- Values and Virtues: Teaching and emphasizing the importance of core values such as honesty, integrity, responsibility, fairness, respect, and empathy. These virtues form the foundation of good character.
- Ethical Decision-Making: Providing individuals with the tools and skills to make ethical decisions in various situations. This involves critical thinking, problem-solving, and considering the consequences of one's actions.
- Social and Emotional Learning (SEL): Incorporating elements of social and emotional education to help individuals understand and manage their emotions, develop empathy, and build positive relationships with others.
- Civic and Global Responsibility: Encouraging a sense of responsibility toward one's community and the larger world. This includes promoting civic engagement, environmental awareness, and a commitment to social justice.
- Resilience and Perseverance: Teaching individuals how to cope with challenges, setbacks, and failures, fostering resilience and perseverance in the face of adversity.
- Positive Behavior Reinforcement: Recognizing and reinforcing positive behaviors to create a positive and supportive learning environment. This can involve praise, rewards, and other forms of positive reinforcement.
- Role Modeling: Demonstrating positive character traits through the actions and behavior of teachers, administrators, and other influential figures in a person's life. Role modeling is a powerful way to convey the importance of good character.

- Community Involvement: Engaging individuals in activities that promote community service and collaboration, fostering a sense of responsibility toward others, and building a strong sense of community.

Character education is often integrated into formal educational curricula and can be implemented in various settings, including schools, families, and community organizations. The ultimate goal is to develop individuals who not only excel academically but also contribute positively to society through their character and ethical conduct.[2]

Instilling Values through Education

The benefits of character education are manifold. Students gain life skills that help them manage relationships and conduct themselves with honor. Schools see improved academic performance and behavior when positive values are instilled. Ultimately, character education aims to develop mature citizens who contribute to a just and caring society. The values we cultivate in our youngest generations will chart the moral course for the future.

Character education actively teaches core ethical and moral values in an intentional, proactive manner. Character education furnishes a more holistic form of schooling that enables students to become healthy, balanced, civic-minded adults. Academic institutions of all types and levels should actively investigate and create caring communities and collaborations with parents to support character development.

Eleven Principles of Effective Character Education

Character education is a broad and evolving field, and there are various principles and frameworks that educators and researchers use to guide character education programs. One widely recognized framework is the "Eleven Principles of Effective Character Education," developed by the Character Education Partnership (now merged with Character.org).

These principles provide a foundation for effective character education programs and actively promote core ethical values like respect and performance values like diligence. They broadly define character beyond just morals.

Many school leaders also use the 11 Principles as a school improvement process. The 11 Principles focus on all aspects of school life, including school culture and climate, SEL, student engagement, and academic achievement, as well as the Multi-Tiered System of Supports (MTSS), Positive Behavioral Interventions and Supports (PBIS), Response to Intervention (RTI), restorative practices, teacher morale, and parent engagement.

MTSS offers a framework for educators to engage in data-based decision-making related to program improvement, high-quality instruction and intervention, SEL, and positive behavioral supports necessary to ensure positive outcomes for districts, schools, teachers, and students.[3]

PBIS is an evidence-based, tiered framework for supporting students' behavioral, academic, social, emotional, and mental health. When implemented with fidelity, PBIS improves social-emotional competence, academic success, and school climate. It also improves teachers' health and well-being. It is a way to create positive, predictable, equitable, and safe learning environments where everyone thrives.[4]

RTI aims to identify struggling students early on and give them the support they need to thrive in school.

PBIS is a specific approach to behavior management, and MTSS is a broader framework that includes academic and behavioral supports, of which PBIS is just one component.

RTI is considered a narrower approach than MTSS. An RTI approach focuses solely on academic assessments, instruction, and interventions. MTSS is a comprehensive framework that includes academic, behavioral, and social-emotional support.[5]

Eleven Principles in Schools

The Eleven Principles in Schools (11 Principles) serve as guideposts for schools to plan, implement, assess, and sustain their comprehensive character development initiative.

The 11 main principles are:

Principle 1: Core values are defined, implemented, and embedded into school culture.

Schools that effectively emphasize character development bring together all stakeholders to consider and agree on specific character strengths that will serve as the school's core values.

Principle 2: The school defines "character" comprehensively to include thinking, feeling, and doing.

The "core values" of a school serve as touchstones that guide and shape a student's thinking, feelings, and actions.

Principle 3: The school uses a comprehensive, intentional, and proactive approach to develop character.

Schools committed to character development look at all they are doing through a character lens. They weave character into every aspect of the school culture.

Principle 4: The school creates a caring community.

A school committed to character development has developed an "ethic of caring" that permeates the entire school.

Principle 5: The school provides students with opportunities for moral action.

Through meaningful experiences and reflection opportunities, schools with a culture of character help students develop their commitment to being honest and trustworthy, volunteering their time and talents to the common good, and, when necessary, showing the courage to stand up for what is right.

Principle 6: The school offers a meaningful and challenging academic curriculum that respects all learners, develops their character, and helps them succeed.

Because students come to school with diverse skills, interests, backgrounds, and learning needs, an academic program that helps all students succeed will be one in which the content and pedagogy engage all learners and meet their individual needs. This means providing a curriculum that is inherently interesting and meaningful to students and teaching in a manner that respects and cares for students as individuals.

Principle 7: The school fosters students' self-motivation.

This principle emphasizes intrinsic motivation over extrinsic motivation. Character means doing the right thing and doing your best work even when no one is looking.

Principle 8: All staff share the responsibility for developing, implementing, and modeling ethical character.

All school staff share the responsibility to ensure that every young person is practicing and developing the character strengths that will enable them to flourish in school, in relationships, in the workplace, and as citizens.

Principle 9: The school's character initiative has shared leadership and long-range support for continuous improvement.

Schools of character have leaders who visibly champion the belief expressed by Martin Luther King Jr. that "intelligence plus character—that is the goal of a true education." These school leaders establish a Character Committee—often composed of staff, parents, community members, and students—and give the Committee the responsibility to design, plan, implement, and assess the school's comprehensive character development initiative.

Principle 10: The school engages families and the community as partners in the character initiative.

Schools of character involve families. Parents are encouraged to reinforce the school's core values at home. School leaders regularly update families about character-inspired goals and activities.

Principle 11: The school assesses its implementation of character education, its culture and climate, and the character growth of students regularly.

Schools of character use a variety of approaches to assess the character development of their students, including student behavior data and surveys. Schools also assess the culture and climate of the school, focusing particular attention on the extent to which the school's core values are being emphasized, modeled, and reinforced.[6]

Importance of Character Development in Individuals

Character education actively prepares students for modern life's challenges. It provides balance to media and Internet messages. As other influences decline, schools actively offer stable community values. Character education principles are traditional yet actively equip students for the future by providing core values with academics to develop responsible citizens.

While academically correct, how does one explain the "what" of character development in terms that K-2 learners will understand?

Figure 1.1 Character education toolbox.[7]

Well, one does so by transferring the principles of character development into meaningful examples that these students are able to relate to.

The concept of character education at the K-2 level is more easily understood by young learners when discussed in age-appropriate terms and with age-appropriate examples.

For example, one approach to discussing the topic of character education with students, ask students to imagine they have a character education toolbox (see Figure 1.1).

Tell students that like tools, in the toolbox, in their character education toolbox, there are things like kindness, sharing, honesty, and being a good friend. Character education is about using these tools to make themselves and the world around them a better place. It's not just about reading and math, but also about learning how to be a good person.

For example, explain that when they share their toys with a friend, that's using their sharing tool from their character education toolbox. Or when they tell the truth, even if it's hard, that's using their honesty tool. These tools help them make friends, solve problems, and be someone that others can trust.

So, character education is like a special set of skills and values that you carry with you every day. It's about being kind, and respectful, and making choices that make you proud. Just like a garden needs care to grow, your character needs attention and practice to become strong and wonderful!

Character development empowers young minds with a moral compass. By instilling values like honesty and responsibility, educators equip students with the tools to make decisions that reflect integrity. These foundational virtues become guiding lights, steering students away from impulsive actions and toward thoughtful, considerate behavior.

Importance of Character Development in Education

One of the great education reformers, Horace Mann, in the 1840s, helped to improve instruction in classrooms across the United States (U.S.), advocating that character development was as important as academics in American schools.

The United States Congress, recognizing the importance of this concept, authorized the Partnerships in Character Education Program in 1994. The No Child Left Behind Act of 2001 renews and re-emphasizes this tradition and substantially expands support for it.

One of the six goals of the Department of Education is to "promote strong character and citizenship among our nation's youth".[8] To reach this goal, the U.S. Department of Education joins with state education agencies and school districts across the U.S. to provide vital leadership and support to implement character education.

To successfully implement character education, schools are encouraged to:

- Take a leadership role to bring the staff, parents, and students together to identify and define the elements of character they want to emphasize.
- Provide training for staff on how to integrate character education into the life and culture of the school.
- Form a vital partnership with parents and the community so that students hear a consistent message about character traits essential for success in school and life.

- Provide opportunities for school leaders, teachers, parents, and community partners to model exemplary character traits and social behaviors.[9]

Educators play a crucial, active role in character building by providing guidance and serving as role models, although parents are the most influential. Schools actively give students opportunities to learn values through peer interactions.

Importance of Character Development for Parents and Communities

Character education actively supports parents' influence rather than supplanting it. Educators and parents should be active partners in instilling values and providing stability amid other influences. Both parents and students actively benefit from character education. It provides students with a moral compass among competing messages. Many parents actively want values-based education.

With work and frequent moves disrupting community ties, schools are actively one of the few stable influences. Character education makes schools actively value-based stable influences. Character education's holistic development of academics, values, and character is critically essential for the next generation. It actively complements parents' roles comprehensively.

Character building in adolescents is insignificantly influenced by parenting. Parents, through parenting, will shape the character of the child. In line with their development and age, children become teenagers and will expand their socialization. As a result, their psychosocial life is also developed. This happens because the scope of their association influences the psychosocial development of adolescents.[10]

The Foundation of Character-Core Values

In the pursuit of building a foundation of character, a set of 14 core values (see Figure 1.2) emerges as the guiding principles that shape individuals into compassionate and responsible members of society.

At the heart of this philosophy lies the value of caring, promoting an innate sense of concern for the well-being of others. Citizenship, an acknowledgment of one's role in a broader community, intertwines seamlessly with the understanding of the role of family. Recognizing

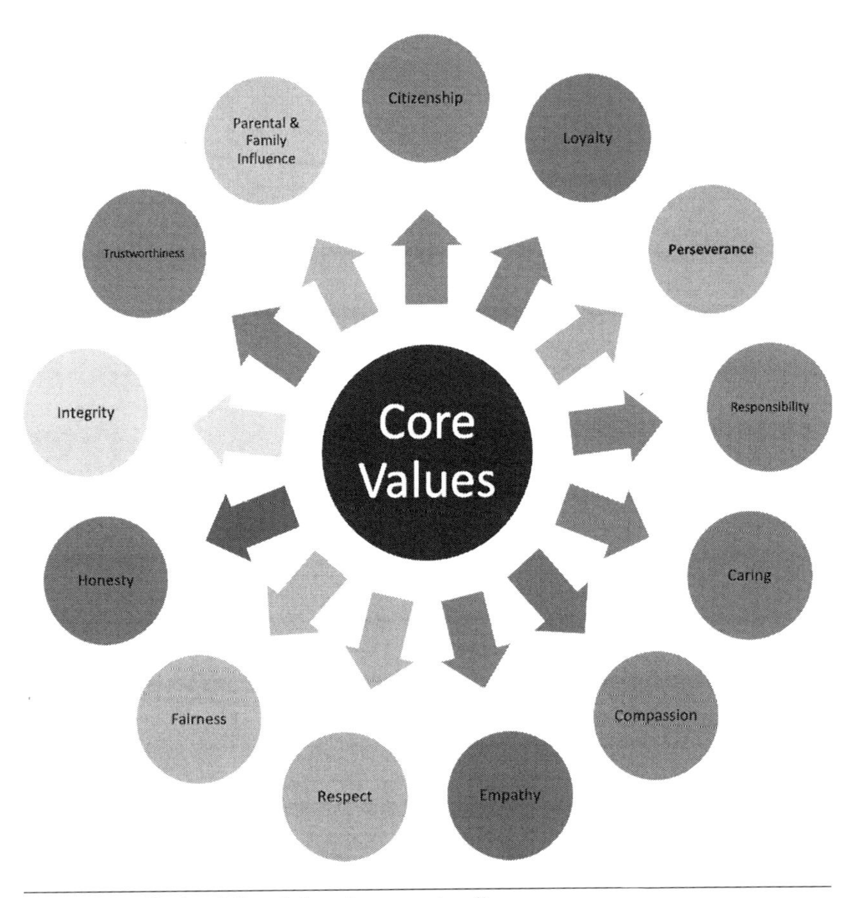

Figure 1.2 The foundation of character-core values.[11]

the profound impact of upbringing and Parental Influence on character development. Compassion and Empathy, cornerstones of human connection, emphasize the importance of understanding and sharing in the experiences of others, fostering a sense of unity and interconnectedness.

Fairness, Honesty, and Integrity form the ethical triad, emphasizing the importance of moral rectitude in all aspects of life. Loyalty, an enduring commitment to relationships and shared values, joins individuals together in a bond of trust and mutual support. Parental Influence, as a separate entity, underscores the pivotal role that family plays in shaping character, serving as a backpack of values and principles that individuals carry into the world. Perseverance, an unwavering commitment to overcoming challenges, stands as a testament to the strength and resilience inherent in an individual's character.

Respect, an acknowledgment of the intrinsic worth of each person, creates an environment of dignity and understanding. Responsibility, both to oneself and to others, manifests as a commitment to contribute positively to society. Trustworthiness, the bedrock of any meaningful relationship, underscores the importance of reliability and honesty in fostering enduring connections.

Together, these 14 core values coalesce to form a solid foundation upon which individuals can build a character marked by empathy, integrity, and a commitment to the well-being of the larger community.[12]

Behavior — What Is Acceptable/Unacceptable?

A good character determines good behavior.

Behavior is something that a person does that can be observed, measured, and repeated. When we clearly define behavior, we specifically describe actions (e.g., Sam talks during class instruction). We do not refer to personal motivation, internal processes, or feelings (e.g., Sam talks during class instruction to get attention).[13]

In a workplace environment, acceptable behavior may be defined as those actions, demonstrations, and language that are supportive of a positive work atmosphere, promote courteous communication, and build teamwork among co-workers.

In a K-2 school setting, defining what behavior is in a way that will be understood by the intended audience and providing a straightforward definition often works best. Behavior, then, to a K-2 student simply refers to how one conducts themselves. It is their actions, reactions, and functioning in response to everyday environments and situations.[14]

Ultimately, there may be many different definitions of behavior, which ones, if any, turn out to be broadly accurate enough for long-term application? Obtaining the answer may better be realized through a question … "What is an accurate set of conditions or criteria that determine what counts as behavior?" Obtaining or even attempting to obtain an answer to that question exceeds the boundaries of this book.

In summary, behavior is simply how one acts.

Individuals, adults, and children can act in a way that their behavior is acceptable and, conversely, in a manner that is perceived by social norms as being unacceptable.

What Rules Define Acceptable Behavior?

Social norms are the perceived informal, mostly unwritten, rules that define acceptable and appropriate actions within a given group or community, thus guiding human behavior. Social norms are learned and accepted from an early age, often in infancy, and held in place by social sanctions ('punishments') for non-adherence to the norm and social benefits ('rewards') for adherence.

A social norm exists when individuals practice a behavior because they believe that others like them or in their community practice the behavior, or because they believe that those who matter to them approve of them practicing the behavior.[15]

Acceptable K-2 Student Behavior

This list is not exhaustive, and, in some cases, other types of behavior deemed acceptable in one school environment, in one society, or in one country may be considered unacceptable in another.

Some acceptable student behaviors include, but are not limited to:

Respectful Communication:

- Using polite language, raising hands to speak, and taking turns during class discussions.
- Raise your hand to respond.
- Listen and follow directions.
- Use an appropriate voice level.

Active Listening:

- Paying attention to teachers, following instructions, and responding appropriately.
- Listen with your ears and your eyes.
- Respect and listen to your classmates.

Sharing and Cooperation:

- Sharing materials, taking turns, and working collaboratively with classmates.

Following Classroom Rules:

- Adhering to established rules and routines within the classroom environment.
- Use technology appropriately.

Completing Assignments:

- Completing classwork and homework assignments to the best of their ability.
- Be prepared with daily assignments and supplies.

Empathy:

- Showing understanding and compassion toward classmates and expressing concern for others.
- Use good manners.
- Use words to solve problems.

Responsible Behavior:

- Taking care of personal belongings and respecting school property.
- Clean up after yourself.

Problem-Solving:

- Engaging in constructive problem-solving and conflict resolution with peers.

Participation:

- Actively participating in class activities, discussions, and group projects.

Respecting Personal Space:

- Understanding and respecting the personal space and boundaries of classmates and teachers.
- Keep your hands and feet to yourself.

Encouraging and reinforcing positive behavior in these areas helps create a positive and supportive learning environment for young children.

Teachers and parents can work collaboratively to model, teach, and reinforce these behaviors to help children develop social and emotional skills.[16]

Unacceptable K-2 Student Behavior

Unacceptable behavior also has moral and societal norms...actions, verbal or physical that are deemed unacceptable as with the previous list, this list is not exhaustive. However, in most cases, types of behavior deemed unacceptable in one school environment, in one society, or in one country may also be considered unacceptable in another.

Some unacceptable student behaviors include, but are not limited to:

Physical Aggression:

- Hitting, kicking, or engaging in any form of physical violence toward peers or adults.
- Angry, aggressive communication (verbal or written).

Bullying:

- Intimidating or harassing classmates, whether verbally or through exclusionary behaviors.

Disruptive Conduct:

- Consistently interrupting lessons, talking out of turn, or engaging in activities that disrupt the learning environment.
- An uttered threat to harm another or damage property.

Defiance of Authority:

- Openly challenging or refusing to follow instructions from teachers or school staff.
- The child who talks back.

Dishonesty:

- Lying about completed assignments, academic performance, or any other aspect of school life.

Destruction of Property:

- Purposefully damaging school property, materials, or belongings of classmates.

Inappropriate Language:

- Use of offensive or disrespectful language toward teachers, staff, or fellow students.

Failure to Follow Rules:

- Ignoring established classroom and school rules consistently.

Stealing:

- Taking items that do not belong to them without permission, whether from classmates or school property.

Inappropriate Touching:

- Engaging in unwelcome physical contact with peers, such as touching inappropriately or invading personal space.

It's important to note that children in grades K-2 are still developing social and emotional skills, and their behavior often requires guidance and teaching rather than punitive measures.

In addressing unacceptable behavior, a proactive and supportive approach that includes communication with parents, counseling, and age-appropriate consequences is generally more effective than a solely punitive measures.[17]

A Student's Character Education Credo

- Character means doing the right thing. When I am at school or at home, I should be honest, kind, fair, and responsible. This helps me be a good friend and classmate. My teachers say developing character is very important.
- Being honest means telling the truth. I will not lie or cheat on schoolwork. Telling the truth even when I make a mistake shows I have integrity. Honesty makes people trust me.
- I also want to be kind. I treat others nicely and help my friends when they need it. Being kind makes my classmates and me feel happy. We get along better when we are kind.
- It's important to be fair, too. I take turns and share with others. When there is a problem, I listen first before reacting. Being fair means caring about how others feel.
- Finally, I want to be responsible. That means doing my part, cleaning up, and taking care of my belongings. Responsibility gives me confidence. My family and teachers are proud when I am responsible. Good character helps me become my best self.

Final Thoughts on Character Education

In the vibrant world of kindergarten through second grade, the journey of education extends far beyond the confines of textbooks and pencils. At this formative stage, character development plays a pivotal role in shaping the very foundation of who we are. As young learners navigate the corridors of knowledge, the importance of fostering positive values and virtues becomes paramount.

Attention at this early stage of a student's academic journey to both creating and fostering a positive and enduring character establishes the foundation for future, long-term confident decision-making skills, constructive life choices, and proactive actions to engage safely in an ever-evolving cyber society.

Cyberbullying and Young Learners

Cyberbullying has become a technology-fed social problem that many families, communities, schools, and other youth-serving organizations address daily. One in seven students in grades K-12 is either a bully or has been a victim of bullying.[18]

To get a better idea of technology's role in bullying, Comparitech surveyed more than 1,000 parents of children over the age of five and asked about their children's cyberbullying experiences. Survey findings indicated that 47.7% of six- to ten-year-olds and over half of those over the age of 11 have experienced bullying in one form or another.[19]

Tween Cyberbullying In 2020, a report from the Cyberbullying Research Center, disclosed that one in five (20.9%) tweens (9–12 years old) has been cyberbullied, cyberbullied others, or seen cyberbullying.[20]

The 2023 Cyberbullying Research Center report, which surveyed a national U.S. sample of approximately 5,000 13- to 17-year-old middle and high school students, indicates that the percentage of students who experienced bullying at school in 2023 remained remarkably similar to levels observed in 2021 (22.6% and 25%, respectively). The report also disclosed that 26.5% of students responding said they had experienced cyberbullying within the past 30 days.[21]

Kids of all ages become victims of cyberbullying, but the risk increases as they grow older.

The following section provides a deeper examination of cyberbullying and provides actions that educators can adopt to address this growing problem with young learners as they begin their introduction to and exposure to technology. Preparing students, even in the K-2 grades, to be aware of and learn how to respond to cyberbullying is a proactive step toward fostering individual self-confidence and cyber-safety practices.

Anti-Bullying

What Is Bullying?

The Anti-Bullying Alliance and its members have an agreed-upon shared definition of bullying based on research from across the world over the last 30 years.

> The repetitive, intentional hurting of one person or group by another person or group, where the relationship involves an imbalance of power. Bullying can be physical, verbal, or psychological. It can happen face-to-face or online.[22]

School bullying takes on many forms and may take place in various locations; however, the initial relationship between parties was formed in a school setting. Bullying may also be in the form of direct, in-person contact, whereas it is more verbal or indirect contact, such as cyberbullying. It is important to ensure that students understand that teachers, counselors, and all support systems within the school, or agency, are available for help and counseling regardless of where the bullying activity takes place, making school a safe environment.

Bullying impacts not only the ability to focus and learn but also physical and mental health, both while the bullying activity is taking place and in the long term.

The Centers for Disease Control and Prevention (CDC) categorizes bullying as a form of youth violence and an adverse childhood experience. The CDC further indicates that bullying is a frequent discipline problem with nearly 14% of public schools reporting that bullying is a discipline problem occurring daily or at least once a week and cites the following statistics.

- Reports of bullying are highest in middle schools (28%), followed by high schools (16%), combined schools (12%), and primary schools (9%).
- Reports of cyberbullying are highest in middle schools (33%), followed by high schools (30%), combined schools (20%), and primary schools (5%).[23]

So prevalent, all 50 United States and many U.S. Commonwealths and Territories have anti-bullying laws and/or policies.[24]

The following discussion and the lesson plans included in Chapter 1 include the following anti-bullying types and definitions and are not an all-inclusive list of anti-bullying types. This is a compilation of bullying types from the University of the People[25] and the Preventing and Promoting Relationships & Eliminating Violence Network (PREVNet).[26]

Although many of the types of bullying listed may appear to be direct, in-person bullying, they may also take place indirectly online.

Cyberbullying: Cyberbullying is any type of bullying that happens online. It can be hurtful comments on a personal site or deceptive private messaging. Includes the use of email, cell phones, text messages, and Internet sites to threaten, harass, embarrass, socially exclude, or damage reputations and friendships.

Physical Bullying: Physical bullying always involves physical contact with the other person. This can mean hand-to-hand, but it can also mean throwing items, tripping, or eliciting others to cause physical harm to a person. This type of bullying can also cause harm to property and belongings.

Verbal Bullying: Verbal bullying means using any form of language to cause the other person distress. Examples include using profanities, hurtful language, negatively commenting on a person's appearance, using derogatory terms, or teasing.

Emotional Bullying: Emotional bullying involves using ways to cause emotional hurt to another person. This can include saying or writing hurtful things, causing others to gang up on an individual, and purposely ignoring, or spreading rumors.

Social Bullying: This includes rolling your eyes or turning away from someone, excluding others from the group, getting others to ignore or exclude, gossiping or spreading rumors, setting others up to look foolish, and damaging reputations and friendships.

Personal Bullying: Personal bullying refers to any sort of bullying, done in any manner that is related to a person's gender or sexuality. Including leaving someone out; treating them badly or making them feel uncomfortable because of their sex; making sexist comments or jokes; touching, pinching, or grabbing someone in a sexual way; making crude comments about someone's sexual behavior or orientation; or spreading a sexual rumor.

Racial Bullying: This includes treating people badly because of their racial or ethnic background, saying bad things about a cultural background, calling someone racist names, or telling racist jokes.

Religious Bullying: This includes treating people badly because of their religious background or beliefs, making negative comments about a religious background or belief, calling someone names, or telling jokes based on his or her religious beliefs to hurt them.

Disability Bullying: This includes leaving someone out or treating them badly because of a disability, making someone feel uncomfortable, or making jokes to hurt someone because of a disability. On a physical level, it includes harmful actions such as blocking ramps and elevators, tripping, or tampering with accessible equipment.

The following discussion focuses specifically on five types of bullying, and how each type may affect your K-2 students. Corresponding lesson plans for classroom exercises and discussion of each of these five bullying types are included in a separate lesson plan section of this book.

Cyberbullying: Cyberbullying is any type of bullying that happens online or via any electronic means, such as cell phones.

Physical Bullying: Physical bullying always involves physical contact with the other person. This type of bullying can also cause harm to property and belongings.

Verbal Bullying: Verbal bullying means using any form of language to cause the other person distress, including teasing.
Emotional Bullying: Emotional bullying involves using ways to cause emotional hurt to another person.
Disability Bullying: Includes leaving someone out or treating them badly because of a disability.

The CDC indicates that different types of violence are connected and often share root causes. Bullying is linked to other forms of violence through shared risk and protective factors. Addressing and preventing one form of violence may have an impact on preventing other forms of violence.[27]

School bullying is a global concern. As such, UNESCO Member States declared the first Thursday of November the International Day against Violence and Bullying at School, including Cyberbullying, recognizing that school-related violence in all its forms is an infringement of children and adolescents' rights to education and their health and well-being.[28]

This presents a great opportunity for schools and child-serving organizations to create an activity or learning fair around this day for student, family, and community awareness. UNESCO themes each year, and an activity or event may be created around the global theme. By incorporating the theme, students may create poster boards and drawings depicting bullying to display on school premises, such as halls, or at school/community events.

As the CDC indicates that bullying leads to other forms of violence, addressing bullying in the classroom, especially in formative years, is critical. Yet a balance needs to be struck in addressing the emotions and reactions of the bully and the child bullied (Figure 1.3).

As is often the case, many bullying responses include "I was just teasing" or "Don't make a big deal of it, they were just teasing." There is a fine line between teasing and bullying, and one that may be easily crossed if not managed properly.

Teasing and Bullying

Sometimes teasing is harmless and playful. Other times, it can be used to hurt others. Even playful teasing can hit raw nerves or be

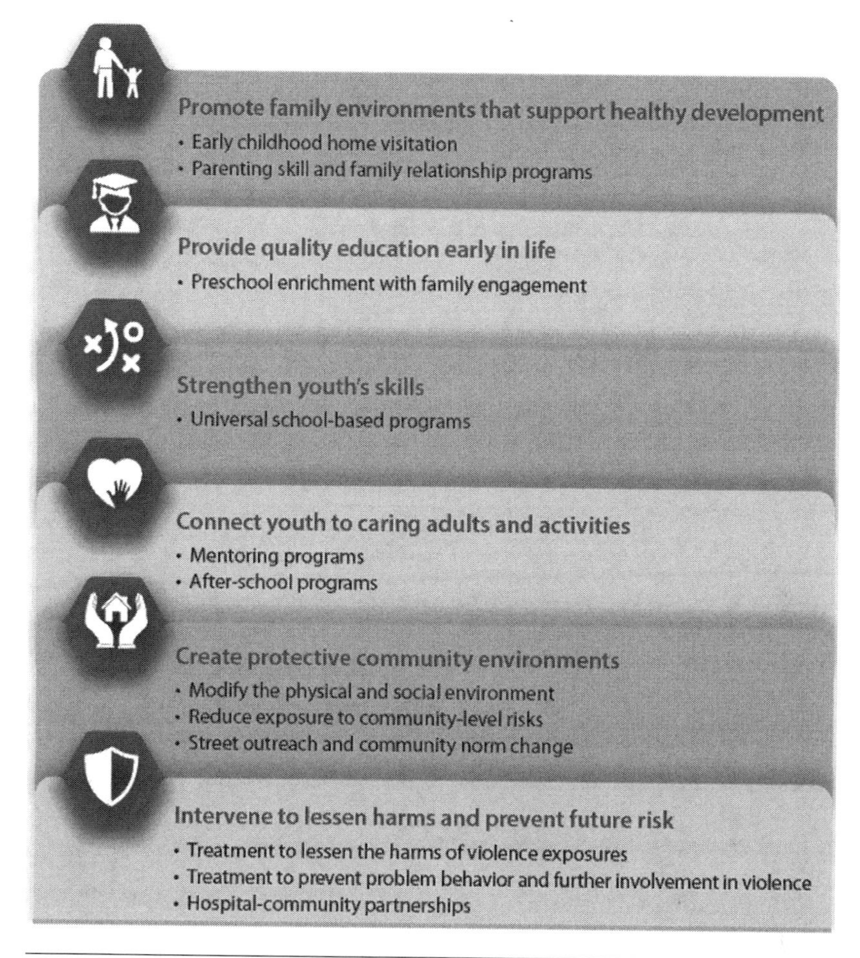

Figure 1.3 CDC: How can we prevent bullying?[29]

misinterpreted, especially when kids struggle with social skills. Many kids tease each other to bond or form relationships. The teasing shows each other they can joke around and still be friends.

But teasing can also be used to communicate the negative. It's often used to establish "top dog" among kids. Also, what's playful to one child may not feel playful to another. In those cases, teasing can lead to hurt feelings.

With these negatives, why not discourage teasing completely? Like any communication, teasing has its purpose. Some topics that are awkward to raise in serious conversation are easier to raise through teasing.

- Teasing and bullying are different.
- Not all teasing is bad. Sometimes it's playful and helps kids bond.
- When teasing is meant to hurt and is done over and over, it can become bullying.

Teasing can also be fun. Think, for example, of the back-and-forth banter that happens in any romantic comedy.

Verbal bullying is different from teasing. It's not done to make friends or to relate to someone. Just the opposite: The goal is to embarrass the victim and make the bully look better and stronger.

The tricky thing is that bullying may start as teasing. But when it's done over and over and is meant to be hurtful or threatening, it becomes bullying.

Unlike kids who are being bullied, kids who are being teased can influence whether it continues or ends. If they get upset, the teaser usually stops.

Sometimes, kids who are trying to tease end up bullying. For example, a child may say something mean-spirited to another, thinking it's playful. This can lead to an argument. Or a child may react angrily to a friendly comment, which may cause other kids to keep their distance.

To address these struggles, it's important to teach kids about the rules of conversation. Help kids sort out when teasing is okay and when it becomes hurtful or borders on bullying. One way to do this is by role-playing with them. This lets kids practice a situation where they get teased, don't like it, and need to respond.

Questions to Ask Kids about Teasing

Maybe you've heard that kids are teasing your child or your student at school. You can ask a few questions to see whether it's good-natured or harmful:

- Are the kids who tease you your friends?
- Do you like it when they tease you?
- Do you tease them back?
- If you told them to stop teasing, would they?
- If you told them that they hurt your feelings, would they say they were sorry?

If the answer to any of these questions is "no" or "I don't know," then it may be a case of negative teasing or even bullying. It's important to find out more.[30]

Judgment and decision-making are integral parts of discussing character education. As indicated in the introduction to this chapter, "The core mission of character education is to guide students to make thoughtful choices and act with integrity." Creating an environment where students feel safe making choices empowers them to feel confident in making future choices.

In the following section, we take a closer look at what good judgment is and how we communicate this principle and this concept to young learners. Using good judgment is not only an important life skill but also becomes especially important when navigating the often confusing and perilous world of technology. Using good judgment is a primary cyber-safety skill!

Using Good Judgment by Making Good Decisions

What Is Good Judgment?

To consider the character development of good judgment in children, guiding children in how to make good decisions is the recommended starting point, and at the earliest ages, such as K-2, this is about making good choices. Good judgment starts with knowing how to access options and the consequences involved in making a choice.

Using good judgment as well as good decision-making is the foundation for ethical and moral decisions the child will continue to make as they get older. The Menninger Clinic, a mental health facility in Houston, states that

> graduates get into the most elite colleges but can't handle college life emotionally—and so take a medical leave of absence and come to the Menninger Clinic for treatment and went on to say that all of these students had had insufficient experience making decisions for themselves, handling setbacks, and managing life's temptations independently.

Authors Johnson and Stixrud continue reinforcing the need for early decision-making skills, stating...

> So, while our impulse is to try to keep our kids safe, our priority as they get older should shift to helping them develop the skills—including the

> decision-making skills—they need to keep themselves safe, along with the confidence that they can trust their judgment and solve problems as they arise. In our view, the time to start supporting kids in making their own decisions isn't when we send them to college—it's when they're young—because we want them to have a ton of practice making and learning from their own decisions...[31]

Making good decisions is about choices. It is easier to make bigger and better decisions when we have the foundation for identifying how to make good choices. Part of helping children make a choice is recognizing that the adult—parent, guardian, or teacher, for example—will need to release control of the decision-making process.

Neuroscience News proposes that children as young as six factor in past choices when making moral judgments. Involving children aged four to nine, the study revealed that younger children's judgments were mainly influenced by the actual outcome, whereas older children factored in what could have been done differently.

Through this counterfactual thinking, children as young as six began to exhibit more nuanced and mature moral assessments. The findings could help guide more effective moral education for young children.

Key Facts:

1. From the age of six, children begin to incorporate past choices into their moral judgments, exhibiting a more mature and nuanced understanding of behavior.
2. In contrast, four- and five-year-old children's moral judgments are influenced solely by the actual outcome.
3. This research provides the first direct evidence that children consider counterfactuals in their moral judgments, paving the way for more comprehensive moral education strategies.

A new study published in the journal *Child Development* by researchers at Boston College in Massachusetts (USA) and the University of Queensland in Australia explores whether four- to nine-year-old children consider past choices when making moral judgments of others. Across two studies, 236 (142 females) children aged four to nine were told stories about two characters who had a choice that led to a good or bad outcome and two characters who had no choice over a good or bad outcome.

The findings showed that from the age of six, children considered what characters could have done when making judgments of how nice or mean they were behaving and that four- and five-year-olds' moral judgments were influenced only by the actual outcome.[32]

Choices — Where to Start?

Many choices are more important than others in terms of impact and risk, so start with less serious scenarios and alternatives. As long as the choice made is not inappropriate and is not harmful to the child or to others, start with simple exercises. It is best to start by presenting the student with a limited number of choices rather than open-ended options. As experience and confidence grow by making these less impactful decisions, the student will be able to tackle more complex choices.

Here are a few examples where both the impact and the risk involved in the choice are low. For parents and guardians, a good place to start is to let children choose what to wear to a party or a family dinner. Select two outfits and let the child pick which to wear. In the classroom, this may extend to play clothes or costumes for a class skit. For educators in the earliest grades, give the child choices about solitary activities such as "Do you want to color a picture, or do you want to look at a picture book?" Both are appropriate, and neither is harmful to the student or other students.

This will take time, as it is essentially an individualized activity! If class sizes are large and this level of one-on-one is not available, options may be to select one child each day or week, or alphabetically by name, as determined by the frequency available. When rotation through the class roster is completed, begin again. The entire class will learn by observing the process as each child makes their decision. Remember, the student is being presented with two right choices, so there is no wrong answer.

A significant result of decision-making and choices is outcomes and consequences. These can be good or bad, or while using the simplest of scenarios, the choice may have little impact. Selecting either drawing a picture or reading a picture book is neither good nor bad. Yet even at the earliest ages, it remains important to discuss consequences. As choices are presented, the outcome of each option must be considered and reviewed with the student.

Figure 1.4 Making good decisions is more than flipping a coin.[33]

Helping the student make a choice should not be left to a flip of the coin (Figure 1.4).

Here are some tips from the Wellspring Center for Prevention to help children develop healthy decision-making habits:
Encourage Critical Thinking:

- Teach children to think critically by asking open-ended questions and encouraging them to weigh the pros and cons of their choices.

Model Good Decision-Making:

- Children learn by example, so be a good role model by making thoughtful decisions and explaining your reasoning to them.

Teach Problem-Solving Skills:

- Help children learn problem-solving skills by encouraging them to identify and evaluate different solutions to a problem.

Foster Independence:

- Allow children to make decisions for themselves, within reason, and encourage them to take responsibility for their choices.

Provide Information:

- Provide children with accurate and age-appropriate information to help them make informed decisions.

Support Resilience:

- Encourage children to bounce back from mistakes and setbacks, and teach them that failures are opportunities to learn and grow.[34]

Do Adults Make Decisions Differently than Children?

Neuroscience News continues to summarize the study, indicating that when making moral judgments about past actions, adults often think counterfactually about what could have been done differently.

"Our findings highlight how understanding the choices someone had is an essential feature of making mature and nuanced moral judgments," says Shalini Gautam, a postdoctoral researcher at Boston College.

> It shows that children become able to do this from the age of six. Children younger than six may not yet be incorporating the choices someone had available to them when judging their actions.[35]

The next section of this book examines the rapid integration and impact of artificial intelligence (AI) on the global academic education process, our individual classrooms, education lesson plans, and discussions with even our youngest learners.

The objectives of this following section are to provide the reader with a foundation, a working literacy, and an understanding of the evolving and oftentimes technical world of AI.

Artificial Intelligence and the K-2 Student

Introduction

The following discussion provides the reader with a non-technical overview of AI…a technology that has become both a bane and a benefit to teachers since the debut of the most talked about, publicly available AI tool, ChatGPT.

The power and popularity of current and emerging AI tools create an environment that is productive, rewarding, and beneficial to users, while also presenting these same users with unseen risks.

Before we "look under the hood" and talk more about AI, discussing this topic with your students may best begin by simply helping your students understand that machines and technology are a part of our daily lives.

Ask your students, for example, about the devices they use (computers, tablets, etc.) and talk about how technology helps us in various ways.

Defining and explaining technology for K-2 students may at times prove challenging, however, it can also be fun while also being instructive (Figure 1.5).

What Is Artificial Intelligence (AI)?

One of the challenges with defining AI is that if you ask 10 different academics, data scientists, or industry practitioners, you will receive

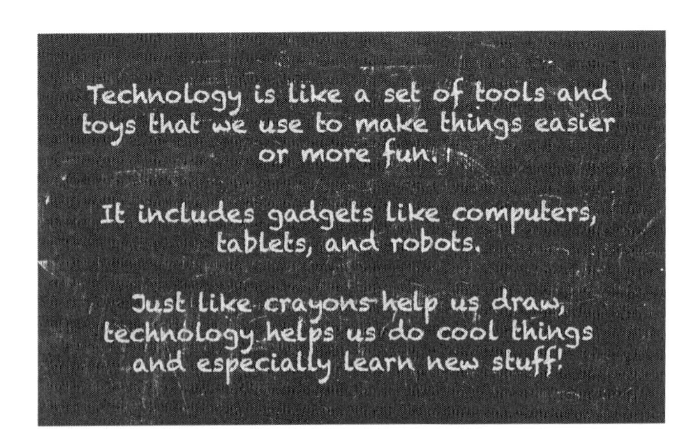

Figure 1.5 Technology defined for K-2 students.[36]

12 different definitions. Defining AI is a moving target. The folks involved with AI today (which seems like almost everyone) haven't converged yet on exactly what a comprehensive, workable, acceptable definition is.

Referring to the 2019 American AI Initiative, Executive Order 13,859, which resulted in the National AI Initiative Act of 2020, Chuck Romine, Director of Information Technology Lab, National Institute of Technology (NIST), provides us with a down-to-earth definition, selected by the authors for its directness, simplicity, and usability when discussing AI with students.

Director Chuck Romine defines an AI system as a system that exhibits its reasoning and performs some sort of automated decision-making without the interference of a human.[37]

How Do AI Systems "Learn?"

Just like developing good characteristics in students throughout their educational journey, technology like AI must be taught to demonstrate good characteristics. AI systems that exhibit characteristics like resilience, integrity, reliability, security, robustness, interoperability, and privacy are exhibiting good characteristics. If AI systems are going to be trusted, useful, adopted, and accepted by people without fear, they must exhibit good characteristics and acceptable behavior.

Teaching AI systems to behave acceptably and embrace approved societal good behaviors and norms, like students, must be taught. Teaching AI systems is not too different from teaching students. In the classroom and through coaching, exercises, and lessons, students are presented with examples of desirable or good characteristics and acceptable behavior. These lessons are repeated throughout a student's academic experience.

Training AI Systems

Training AI systems involves a process similar to teaching a computer how to perform specific tasks or make decisions by exposing it to large amounts of data. This process is known as machine learning (ML). At its core, ML is the subset of AI that enables systems to learn and improve from experience without being explicitly programmed.

The fundamental concept is to input millions, if not billions, of bytes of data that represent information into the AI system. This input or "feeding" process allows the AI system to recognize patterns, make predictions, or classify information autonomously.

One primary method of training AI is called supervised learning. In this approach, the AI system is provided with a labeled dataset, where each input is paired with the correct output. For instance, if we want the AI to identify pictures of cows and horses, we will present the AI system with a collection of images labeled as either a cow or a horse. The AI learns by adjusting its internal parameters through repeated exposure to these labeled examples, refining its ability to make accurate predictions. Supervised learning is analogous to a teacher guiding a student with correct answers during a learning process.

Another key training method is unsupervised learning. Here, the AI explores data without explicit guidance, seeking to identify hidden patterns or relationships. Clustering, a common unsupervised learning technique, involves grouping similar data points together. Imagine a scenario where the AI receives a mix of unlabeled pictures and independently discovers that some share characteristics, effectively categorizing them into groups. This unsupervised learning is comparable to a student independently organizing information based on inherent similarities.

Reinforcement learning is a third approach, inspired by behavioral psychology. In this method, the AI, often referred to as an agent, learns to make decisions by interacting with an environment. It receives feedback in the form of rewards or penalties based on the actions it takes. This trial-and-error process enables the AI to learn optimal strategies for maximizing rewards. Reinforcement learning can be compared to a student learning through experimentation and adjusting their behavior based on outcomes.

The training process also involves neural networks, which are computational models inspired by the human brain's structure. Neural networks consist of interconnected nodes (or artificial neurons) organized into layers. During training, the network adjusts the weights assigned to connections between nodes, optimizing its ability to make accurate predictions. The deep learning paradigm involves neural networks with many layers, allowing them to capture intricate features in

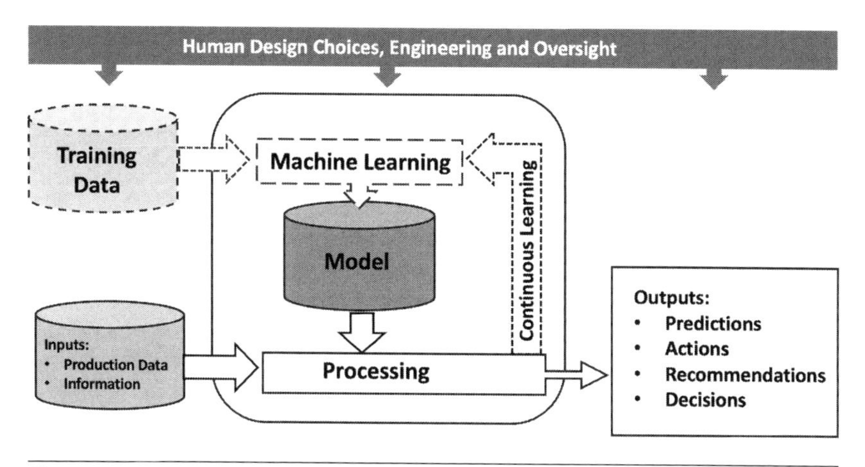

Figure 1.6 AI functional overview.[38]

the data. Visualize this as a student's brain adapting and strengthening connections as they acquire more knowledge and skills over time.

To sum up our review of AI and the various subsets that make up the broader field of AI, we are left with one elementary question… "How does artificial intelligence use data to make informed decisions?"

The answer is uncomplicated and straightforward…AI uses data to make informed decisions through a process called "machine learning." ML is a subset of AI that enables systems to learn from data, identify patterns, take actions, and make recommendations, predictions, or decisions based on that data (see Figure 1.6).

An Example for the K-2 Student

So, how do we translate this information on how AI is trained to provide us with answers to our questions into words and concepts that are understandable to K-2 students?

Here is one example of an in-class exercise you may wish to consider and adapt for your learners.

Training AI systems is like teaching a robot how to do something. Begin your discussion on how AI "learns" by asking your students to imagine that they have a new friend who is a robot.

> The robot's name is GiggleGear and you want to show GiggleGear how to recognize animals like cows and horses. First, you need to collect lots

> of pictures of cows and horses so GiggleGear can learn what they look like. This is like giving GiggleGear a big photo album to study.
>
> Next, you and GiggleGear play a game.
>
> You show GiggleGear a picture and say, "This is a cow," and GiggleGear learns to remember that cows are often black and white or brown, have big, floppy ears, and have long tails with a tuft of hair at the end.
>
> Then, you show another picture and say, "This is a horse," and GiggleGear learns that horses have a long mane of hair on their neck and a fluffy tail, have long legs with hooves at the bottom, and come in different colors like brown, black, white, and sometimes with spots. You keep playing this game with many pictures until GiggleGear knows how to tell the difference between cows and horses all by itself.
>
> Once GiggleGear learns a lot, it can recognize cows and horses in new pictures it has never seen before! Training AI is like teaching your robot friend GiggleGear to understand and figure things out on its own by looking at lots of pictures and learning from them.

In summary, training AI systems encompasses various methods, such as supervised learning with labeled datasets, unsupervised learning for discovering patterns independently, and reinforcement learning for decision-making through interaction with an environment. Neural networks, inspired by the human brain, play a crucial role in adjusting their connections to improve accuracy. The process is analogous to guiding a student through examples, allowing them to learn independently, and refining their decision-making through experience.

What Is Generative AI (GEN-AI)?

The popular GEN-AI tool, ChatGPT 3.5, was presented with the following request… "Please explain what generative AI is using non-technical terms." ChatGPT 3.5 responded by providing this definition….

> Generative AI is like a creative virtual artist. It's a type of artificial intelligence that can generate new content, such as images, text, or even music, based on patterns it learned from existing data. Instead of just following specific instructions, it can come up with original ideas by understanding and mimicking the style or structure of the data it was

> trained on. It's a bit like a smart assistant that can create things on its own, making it useful for various tasks like generating realistic images or writing creative text.

Asked to explain further, ChatGPT 3.5 added the following....

> Generative AI refers to a class of algorithms and models designed to generate new data samples that resemble a given dataset. These models are trained on a set of input data and learn the underlying patterns, structures, and statistical relationships present in that data. Once trained, the generative AI can produce new, similar data points that share the characteristics of the original dataset.
>
> In essence, generative AI uses statistical methods to understand and reproduce the patterns observed in the training data, enabling it to create novel, synthetic data instances. This capability finds applications in various domains, including image synthesis, text generation, and data augmentation in machine learning.[39]

When explaining GEN-AI to your students, simplicity is best .... GEN-AI is like a smart computer artist who can create new pictures or stories by learning from old ones.

Caution: AI Ahead

One risk associated with GEN-AI is exactly what makes it "intelligent," the use of excessively large amounts of data. Data that can come from both known (substantiated) and unknown (unsubstantiated) sources.

Due to the vast amounts of data that GEN-AI models have access to, the ability to generate either intentionally or unintentionally, harmful content is highly probable. Content that may inaccurately attribute comments to someone who never made the statement or alter images to deceive the viewer into believing that the image represents an authentic scene, person, or event.

In the field of AI and ML, this is referred to as hallucination. The production of plausible-seeming but factually incorrect output by a GEN-AI model that purports to be asserting the real world; however, the AI system provides an answer that is factually incorrect, irrelevant, or nonsensical because of limitations in its training data and architecture.

The potential for GEN-AI models to increase efficiency and productivity, make better decisions, improve the "speed" of business, reduce human error, and improve services comes with ethical concerns and risk exposures inherent in AI technologies and GEN-AI models.... accountability and transparency, the potential to create and distribute harmful content, privacy exposure, loss of data protection, trust and explainability concerns, job displacement, unemployment, economic disruption, and data biases.

Throughout their academic journey, students learn about being an ethical person and that responsibility and trust represent good characteristics and acceptable behavior. These lessons need to be reinforced and discussed with students in terms related to the world of technology. This is even more important now with the advent and presence of AI as a tool, a resource gaining increasing popularity in all sectors of society, including academia, and the increasing availability for student access and use.

Just as students learn what ethical and responsible behavior is, that trust is earned, how to trust, and that not everyone can be trusted, they must be provided with instruction and guidance on using AI ethically and responsibly, and that AI cannot always be trusted, meaning that AI, as it exists currently, will not always provide the correct answer.

Ethics: A General Introduction and Overview

Before we provide a generally accepted definition of ethics, we must examine the core principles that, when working together, lead to this generally accepted definition of ethics.

Core Principle #1: Values. Values are the foundation of an individual's ability to judge between right and wrong; they frame the decisions we make. Values include a deep-rooted system of beliefs that guide a person's decisions. They form a personal, individual foundation that influences a particular person's behavior.[40]

Core Principle #2: Morals. Morals are the principles that guide individual conduct within society, and they emerge out of core values. While morals may change over time, they remain the standards of behavior that we use to judge right and wrong.

In the study of ethics, the terms amoral and immoral are often interchanged; however, in application, they are quite different.

Amoral refers to a lack of moral principles or a disregard for moral values altogether. When someone or something is described as amoral, it means they do not differentiate between right and wrong, nor do they consider the ethical implications of their actions.

Immoral, on the other hand, refers to behavior that is in direct violation of accepted moral principles or standards. Immoral actions are considered wrong or unethical and often lead to negative consequences for others or society as a whole.

Amoral behavior lacks a moral compass altogether, whereas immoral behavior knowingly goes against established moral standards. While amoral actions may not carry inherent moral judgment, immoral actions are generally seen as wrong and ethically objectionable.

Core Principle #3: Judgment. Judgment refers to the process of evaluating and making decisions about the moral rightness or wrongness of an action, behavior, or belief. Ethical judgment involves considering various moral principles, values, and standards to assess whether a particular action is morally permissible, obligatory, or prohibited. Ethical judgment consists not of getting the right answers all the time but.... of consistently asking the right questions.

Core Principle #4: Norms. Norms refer to the established standards, rules, or principles that govern and guide human behavior within a particular social or cultural context. These norms dictate what is considered morally acceptable or unacceptable within a given community, group, or society. Norms serve as a framework for evaluating and judging the ethicality of actions, decisions, and practices. Norms help to promote ethical conduct and help prevent harmful or unethical behaviors.[41]

Core Principle #5: Standards. Standards refer to the specific criteria or benchmarks that are used to assess and measure ethical behavior or the moral quality of actions, decisions, or practices. These standards serve as guidelines for evaluating whether an individual, organization, or society is adhering to ethical principles and behaving in a morally responsible manner. Standards are crucial in promoting ethical behavior

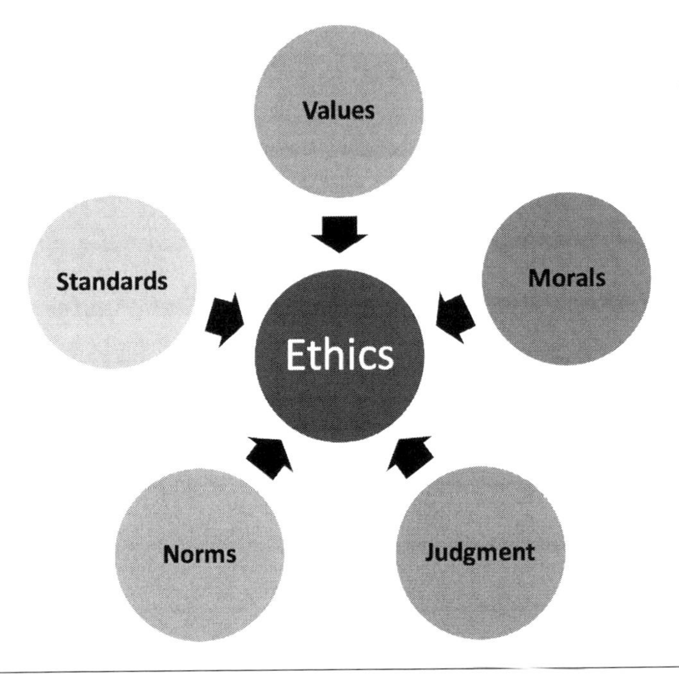

Figure 1.7 Core principles of ethics.[42]

> and holding individuals and institutions accountable for their actions, ensuring that they align with recognized ethical values and principles (see Figure 1.7).

Values are "judgments of worth," moral principles that should have a certain weight in the choice of an action. Morals refers to what is judged as right, just, or good. Judgment is the ability to make accurate determinations. Norms state what is morally correct behavior in a certain situation. Standards assess and measure ethical behavior. Working together, these five core principles lead us to a generally accepted definition of ethics.

Ethics … the collection of values, morals, judgments, norms, and standards, which provides a framework for acting.

In their article "A Framework for Ethical Decision Making," the team from Markkula Center for Applied Ethics at Santa Clara University, states that

> ethics refers to standards and practices that tell us how human beings ought to act in the many situations in which they find themselves—as friends, parents, children, citizens, businesspeople, professionals, and so on. Ethics is also concerned with our character. It requires knowledge, skills, and habits.

The team goes on further in defining ethics by identifying what ethics is not. One characteristic of what ethics is not, and which is receiving much attention and debate, concerning GEN-AI models is… ethics is not science.

> Social and natural science can provide important data to help us make better and more informed ethical choices. But science alone does not tell us what we ought to do. Some things may be scientifically or technologically possible and yet unethical to develop and deploy.[43]

Ethical Use of AI

If the past 24 months are any indicator of the widespread acceptance of GEN-AI as a tool for the masses, students as well as adults must be taught to engage with and use AI ethically.

The ramifications of the unethical use of AI are numerous, and as AI matures, the list of potential consequences of unethical AI grows as well.

Currently, the most significant issues with unethical AI are:

- Privacy Erosion: The unethical use of AI poses a significant threat to privacy. As AI systems gather and analyze vast amounts of personal data, there is a risk of unauthorized access, data breaches, and misuse.
- Bias and Discrimination: Unethical AI practices can perpetuate and even exacerbate societal biases. If the training data used to develop AI algorithms contains biases, the AI systems may produce discriminatory outcomes.
- Security Threats: The unethical use of AI can introduce new security threats as malicious actors leverage AI to enhance the sophistication of cyberattacks.
- Job Displacement and Economic Inequality: The deployment of AI in the workforce without ethical considerations can lead to job displacement and economic inequality.
- Autonomous Weapons and Warfare: The unethical use of AI in the development of autonomous weapons raises profound ethical and humanitarian concerns.

Addressing the most significant issues of unethical AI requires a global recognition of the risks that can be inherent within AI and an effort to establish ethical frameworks, regulations, education programs, and responsible AI practices.

Teach Your Children Well

The message contained within the lyrics of the Crosby, Stills, Nash & Young song "Teach Your Children" is as relevant today as when Graham Nash wrote it in 1970...we all have a responsibility to pass on our values, beliefs, and dreams to the next generation so that they can build a better world.[44]

AI will have a significant impact on the world in which our students live, learn, and grow. Teaching students to use AI ethically and responsibly is preparing them to manage AI as adults and use AI to build a safer, better world.

Teaching K-2 students about ethical behavior and behaving responsibly can at times be challenging. For educators, applying these life skills to AI adds to this challenge.

Educators may wish to consider the following when discussing the ethical and responsible use of AI with their students.

1. Emphasize that AI is designed to help people and make their lives easier. Provide examples such as smart assistants like Siri or Alexa, which can answer questions or play music. Explain that AI is a tool that can assist us in various ways.
2. Teach the importance of privacy by explaining that AI works with information and data. Emphasize that it's essential not to share personal information online or with AI devices without asking a grown-up for permission.
3. Discuss the idea that AI should be fair and treat everyone the same. Use examples like choosing games or activities, highlighting that AI should not favor one person over another.
4. Stress the importance of being truthful when interacting with AI. Explain that providing accurate information helps AI learn and do its job better. Encourage students to use AI responsibly and not to trick it.

5. Explain that AI learns from people, just like students learn from their teachers. Encourage a collaborative mindset by discussing how we can teach AI good things and help it become smarter in positive ways.
6. Introduce the concept of being a good digital citizen. Teach the importance of using AI and technology in a kind and respectful manner. Discuss how to communicate online and treat others with kindness, even when using AI.
7. Emphasize that it's okay to ask for help when using AI. Encourage students to talk to their parents, their teacher, or other grown-ups if they have questions about AI or if something online makes them feel uncomfortable.
8. Clarify that AI has limits and may not always have the right answers. Teach students to have realistic expectations and that it's okay if AI doesn't know everything.
9. Discuss the importance of balancing screen time and offline activities. Teach students that while AI can be fun and helpful, it's essential to have a balance by playing outside, reading books, and spending time with family and friends.

By incorporating these concepts into age-appropriate discussions and activities, the K-2 educator can lay the foundation for ethical and responsible AI use, promoting positive digital behavior among young learners.

Chapter 2, "Nurturing Digital Citizens: Cyber Safety for Early Learners," provides the educator with insights into discussing and teaching students the necessity and importance of cyber safety as young learners and as they grow in an ever-changing, technology-infused digital world.

LESSON PLANS

Kindergarten

Character Education

TOC Title: G-K Character

Lesson Title: What Is Character?

Grade Level: Kindergarten

Duration: 40 minutes

Objective:

- Students will understand and be able to identify the key components of character, i.e., honesty, respect, responsibility, fairness, and compassion.

Suggested Materials:

- Large chart paper or whiteboard.
- Markers.
- Pictures or flashcards depicting various scenarios related to each value.
- Picture books that focus on good character, e.g., The Empty Pot by Demi that teaches the value of being honest, or another book that addresses the topic at an age-appropriate level.

Procedure

Introduction (15 minutes):

a. Class Discussion: Start by introducing the concept of a toolbox. Show students a real toolbox or pictures of one. Explain that just like a toolbox holds tools that help us fix things, we have our own "character toolbox" that helps us be good people.
b. As a group, discuss what it means to be honest, respectful, responsible, fair, and compassionate.
c. Class-friendly definitions: Use the chart paper or whiteboard to write down each value and write definitions in terms students will understand. Encourage students to help create class definitions.
d. Suggested definitions (class definitions need not match word for word):

- Honesty: When someone is truthful in the things they say and do.
- Respect: treating others (and yourself) in a way that shows you care about their well-being and how they feel.
- Responsibility: doing what you are supposed to do and accepting the consequences (good or bad) of your actions.
- Fairness: treating everyone equally and not playing favorites.
- Compassion: treating others with kindness and wanting to help them.

Read Aloud (10 minutes):

a. Suggested reading: The Empty Pot by Demi, or another picture book that focuses on good character.

Class Discussion (10 minutes):

a. Discuss the "tools" students noticed being demonstrated in the story. Ask questions like:
   - "What character tools did you notice being used in the story?"
   - "Tell us about a time you used this tool."

Conclusion (5 minutes):

a. Wrap up by restating the values in the character toolbox learned today and encouraging students to practice these values every day.

Homework (Optional):

a. Encourage students to talk about these values with their family and share a story or example of when they demonstrated one of these values at home.

Assessment:

a. Rather than a formal assessment, a qualitative assessment of each student's understanding and application of the values taught is suggested. Focus on how well students understand the concepts and their ability to apply these values in their lives.

Acceptable Behavior

TOC Title: Grade-K Acceptable Behavior

Lesson Title: Grade-K Acceptable Behavior

Grade Level: Kindergarten

Duration: 30 minutes

Objective:

- Students will understand what acceptable behavior is.
- Students will be able to identify acceptable and unacceptable behavior.

Suggested Materials:

- Large paper or whiteboard.
- Markers.
- Images or flashcards depicting different behaviors (e.g., sharing, listening, raising hands) have some images of unacceptable behavior for an assessment later.
- Stickers or rewards (optional).

Procedure

Introduction (10 minutes):

a. Begin by asking the students what they think "good behavior" means. Encourage them to share examples or experiences.
b. Write down their ideas on the whiteboard or large piece of paper. Use kindergarten language and drawings to illustrate each behavior they share.

Activity (15 minutes):

a. Show images or flashcards depicting various behaviors (e.g., sharing, listening to the teacher, raising hands, using kind words).
b. Have the students identify and describe each behavior. Discuss why it is important and how it contributes to a positive classroom environment.
c. Optional activity: Role-playing – encourage volunteers to role-play these behaviors by acting out scenarios where they demonstrate good behavior.

Conclusion (5 minutes):

a. Summarize the behaviors and their importance in creating a safe and happy classroom environment.
b. Thank the children for their participation. Encourage them to practice these behaviors both in school and at home.

Homework (Optional):

a. Students draw a picture of themselves displaying one of the good behaviors discussed in class. This could be shared with the class during the next session.

Assessment:

a. Mark two areas in the classroom or two containers: "Good Behavior" and "Needs Improvement."
b. Distribute flash cards or images from earlier in the lesson to each student.
c. Have students place their card in the area they think it belongs and state why they think it goes there.

Assessment Criteria:

a. Students can correctly sort behaviors into "Good Behavior" and "Needs Improvement" categories.
b. Students are now able to explain why certain behaviors fall into each category.

Anti-Bullying

TOC Title: Cyberbullying

Lesson Title: Cyberbullying

Grade Level: Kindergarten

Duration: 35 minutes

Objectives:

- Teach kindergarten students what cyberbullying is and how to be kind and respectful online.

Materials:

- Pictures or drawings representing different emotions (happy, sad, angry, etc.).
- Whiteboard or flipchart.
- Markers.
- Picture books about kindness and empathy, such as The Technology Tail by Julia Cook or another book that addresses the topic at an age-appropriate level.

Procedure

Introduction (5 minutes):

a. Ask students what it means to be kind to one another. Discuss examples of kindness.
b. Show pictures of different emotions and ask students to identify each emotion.

(Optional) Read Aloud and Activity (20 minutes):

a. Read a storybook that emphasizes kindness and empathy, like The Technology Tail.
b. Role-playing scenarios: Create different cyberbullying scenarios for the students to act out. Facilitate making good responses to situations, such as telling a trusted adult, or being kind to someone who is being treated poorly online.

Activity – Understanding Cyberbullying (10 minutes):

a. Explain that sometimes when people use computers, tablets, or phones, they can say or do things that hurt others' feelings. This is called cyberbullying.
b. Use scenarios that children can relate to. For example, "Imagine if someone said mean things about your drawing online. How would that make you feel?"
c. Talk about why it is important to be kind, not just in person but also online.

Interactive Activity – Being Kind Online (15 minutes):

a. Draw a simple picture or scenario on the whiteboard or flip-chart (e.g., two people chatting online).
b. Ask the children to suggest kind and unkind things that could be said in that situation. Write these on the board.
c. Emphasize that it is important to always be kind and polite when talking to others, whether it's in person or online.

Conclusion (5 minutes):

a. Summarize the lesson by reminding the students that being kind online means using nice words and not saying things that could hurt someone's feelings.
b. Encourage children to always talk to a trusted adult if they ever feel upset or uncomfortable because of something online.

Assessment:

a. Provide coloring sheets or simple worksheets, and ask children to draw or write about being kind online or how they would help a friend who is feeling sad because of something they saw online.

TOC Title: Kindergarten Physical Bullying Lesson

Lesson Title: Understanding Physical Bullying

Grade Level: Kindergarten

Duration: 45 with read aloud. 30 minutes without.

Objective:

- The students will recognize physical bullying behaviors, understand the impact on others, and cultivate empathy.

Suggested Materials:

- Picture books or visual aids depicting bullying scenarios such as Pushing Isn't Funny: What to Do about Physical Bullying (No More Bullies) by Melissa Higgins, Chester Raccoon and the Big Bad Bully by Audrey Penn, or another book that addresses the topic at an age-appropriate level.
- Images or drawings illustrating different emotions.
- Large paper or whiteboard for drawing and discussion).
- Art supplies (crayons, markers, etc.).
- Assessment sheets with simple questions or activities (e.g., drawing, matching, or identifying emotions).

Procedure

Introduction (10 minutes):

a. Discuss feelings and emotions. Use images or drawings to showcase various emotions (happy, sad, scared, etc.). Ask students to identify these emotions.
b. Introduce the concept of bullying in a child-friendly way. Explain that bullying means repeatedly being mean to someone or hurting someone on purpose.

Read Aloud (10 minutes) [Optional if reading books are not available]

a. Read a story or show visuals depicting a bullying scenario involving physical actions. Ensure it is age-appropriate and highlights the feelings of both the bullied and the bully.

Class Discussion (5 minutes):

a. Talk about the story. Ask questions like:
- "How do you think the characters being bullied felt?"
- "Why do you think it's not okay to hurt others?"
- "What could the bully have done differently?"
- Encourage empathy by asking how they would feel in the victim's place.

Activity (15 minutes):

a. Drawing Emotions: Have students draw different emotions they've discussed.
b. Encourage them to think about how someone might feel if they're bullied or hurt.
c. Discuss these drawings afterward, emphasizing understanding and empathy.

Conclusion (5 minutes):

a. Highlight key points about bullying, empathy, and the importance of being kind to others.
b. Encourage students to be helpers and talk to a trusted adult if they witness or experience bullying.

Assessment:

a. Assessment may involve evaluating student drawings, participation, and completion of activities. Focus on their ability to recognize emotions and demonstrate empathy toward others.

TOC Title: K Verbal Bullying
Lesson Title: K Verbal Bullying
Grade Level: Kindergarten
Duration: 40–50 minutes

Objective:

- Students will understand the impact of hurtful words.
- Students will understand that using kind words helps everyone feel happy and valued.
- Students know how to foster empathy and encourage positive communication using kind words.

Suggested Materials:

- A picture book on kindness and friendship such as Chrysanthemum by Kevin Henkes or another book that addresses the topic at an age-appropriate level.
- Paper.
- Crayons/markers.
- Whiteboard and markers.

Procedure

Introduction (10 minutes):

a. Begin with a discussion about feelings.
b. Ask children how certain words or actions make them feel happy or sad. Use simple examples and encourage sharing.

Read Aloud (10–15 minutes) [Optional if reading books are not available]

a. Choose a picture book that highlights kindness and discusses the importance of using kind words. (Chrysanthemum by Kevin Henkes is suggested, or another book that addresses the topic at an age-appropriate level.)

Activity (10–15 minutes):

a. Divide the children into pairs.
b. Assign roles: one child acts as the "speaker," and the other is the "listener."

c. The speaker will say kind words or compliments to the listener, while the listener practices active listening by making eye contact and nodding.
d. Students switch roles after a few minutes.

Conclusion (10 minutes):

a. Gather the children and discuss what they learned from the activities.
b. Ask questions like:
   - "How did it feel to say kind words?"
   - "How did it feel when someone was saying kind things to you?"
   - "Why is it important to use kind words?"
   - "How can we help someone who feels sad because of unkind words?"
c. Reiterate the lesson's message: "Using kind words helps everyone feel happy and valued."

Assessment:

a. An informal assessment of the speaker & listener activity and discussion afterward can be used.

TOC Title: Emotional Bullying Grade K

Lesson Title: Understanding Emotional Bullying

Grade Level: Kindergarten

Duration: 40–50 minutes

Objective:

- Kindergarten students will understand what emotional bullying is and learn ways to respond and support each other.

Suggested Materials:

- Whiteboard or chart paper.
- Markers.
- Pictures or illustrations depicting emotions.
- Storybooks about kindness and empathy, such as The Invisible Boy by Trudy Ludwig or a similar book.

Procedure

Introduction (10 minutes):

a. Begin by asking students about their feelings. Use pictures or illustrations to show various emotions (happy, sad, angry, etc.). Ask them to identify these emotions.
b. Explain that sometimes people can make others feel sad or upset by saying mean things or excluding them. This is called emotional bullying.
c. Point out that it is important to be kind to others and help if someone is feeling sad or upset.

Read Aloud (10 minutes) [Optional if reading books are not available]

a. Read a storybook that emphasizes kindness and empathy, like The Invisible Boy, or another book that addresses the topic at an age-appropriate level.

Class Discussion (5 minutes):

a. Discuss the story together. Ask questions like:
   - How do you think Brian felt when nobody noticed him?
   - Why is it important to always be kind and include others?
   - How can we help if someone is feeling sad or left out?

Activity (10 minutes):

a. Provide paper and markers/crayons.
b. Students draw a picture of themselves with friends, which shows how they can help each other feel happy and included.
c. Encourage students to share their drawings and describe what is happening in them.

Empathy Activity (10 minutes):

a. Divide the class into small groups.
b. Provide scenarios or role-play situations where a child might feel left out or sad because of something another child said or did.
c. Encourage the groups to come up with ways to help the upset child feel better.
d. Each group presents their solution to the class.

Conclusion (5 minutes):

a. Recap the main points discussed about emotional bullying.
b. Encourage students to remember to be kind and support each other.
c. Remind students that seeking help from teachers or adults is important if someone is being unkind or if they are feeling upset.

Homework (Optional):

a. Ask students to share something nice for a classmate or family member and report back the next day.

Assessment:

a. Informal assessment can be done throughout the lesson by observing students' participation in discussions, their willingness to share their thoughts, and their understanding of the key concepts discussed.

TOC Title: K Disability Bullying Lesson

Lesson Title: Disability Bullying

Grade Level: Kindergarten

Duration: 35–45 minutes

Objective:

- Students will understand what disabilities are.
- Students will be able to promote empathy and kindness toward everyone.

Suggested Materials:

- (Optional) picture book related to disability bullying. Suggestion: We'll Paint the Octopus Red by Stephanie Stuve-Bodeen or another book that addresses the topic at an age-appropriate level.
- Blank paper or coloring sheets.
- Crayons, colored pencils, or markers.

Procedure

Introduction (10–15 minutes):

a. Discuss differences among people. Talk about how some people have different abilities, and some people need help doing things that other people can do on their own.
b. Ask students to name things that they need help with.
c. Define what a disability is. Emphasize that it's a part of who someone is, just like hair color or shoe size.
d. Define bullying and include that sometimes people are targeted by a bully because of their disability.
e. Ask open-ended questions about how we should treat others who might be different from us.

Read Aloud (10 minutes) [Optional if reading books are not available]

a. We'll Paint the Octopus Red by Stephanie Stuve-Bodeen, or a similar book.

Class Discussion (5 minutes):

a. Ask questions about the book to prompt discussion:
   - How did the girl feel when she found out her brother was going to be different?
   - What did the girl learn about her brother's differences?
   - Why is it important to be kind to people who are different from us?

Activity (5–10 minutes):

a. Circle time discussion: Facilitate a safe and supportive environment. Ask students to talk about their feelings, experiences, and concerns about bullying.
   - This activity should be repeated regularly to foster a safe and respectful classroom environment. Integrating it into everyday activities can help instill empathy and positive behavior and promote a safe and respectful classroom environment.

Conclusion (5 minutes):

a. Recap what the children learned about disabilities and kindness.
b. Emphasize the importance of treating everyone with respect, regardless of differences.

Assessment:

a. Informal assessment can be done based on students' participation in class activities/discussions.

Using Good Judgment:

TOC Title: Using Good Judgment

Lesson Title: Using Good Judgment

Grade Level: Kindergarten

Duration: 40 minutes

Objective:

- Introduce the concept of good judgment to kindergarteners and help them understand simple scenarios where good judgment can be applied.

Suggested Materials:

- Storybook: Choose a simple story that illustrates characters making good or poor decisions, e.g., Great Choice, Camille by Stuart J. Murphy or a similar book.
- Pictures or flashcards depicting scenarios where good judgment is required (e.g., crossing the road, sharing toys).
- Large paper or whiteboard and markers.
- Stickers or small rewards.

Procedure

Introduction (5 minutes):

Discussion:

a. Ask students what they think good judgment means. (See Chapter 1 content.)

Read Aloud (10 minutes) [Optional if reading books are not available]

a. Read the selected book aloud, pausing occasionally to ask questions like:
   - "What do you think the character should do in this situation?"
   - "Why do you think that's a good choice?"
   - "What might happen if the character made a different choice?"

Class Discussion (5 minutes):

a. Discuss the choices the character(s) made, whether those choices show good judgment, and how we know if they were good or bad choices.
b. If a story was not read, provide examples of choices people make to which kindergarteners can relate.

Activity (10 minutes):

Art Mini-Project:

a. Provide each child with paper and art supplies. Instruct them to draw a picture of themselves making a good decision, e.g., sharing toys, helping a friend, or being safe when crossing the street.

Conclusion (5 minutes):

a. Invite the children to share their drawings and explain the good decisions they made.

Assessment:

a. To reinforce the concept of using good judgment, frequently look for opportunities to discuss and reward (reinforce) students' use of good judgment throughout the school year.

Artificial Intelligence

TOC Title: Kindergarten AI Lesson (My Robot Friend)

Lesson Title: Kindergarten AI Lesson (possible title – My Robot Friend)

Grade Level: Kindergarten

Duration: 35 minutes

Objective:

- Students will understand the concept of AI.
- Students will understand what it means to use AI responsibly.

Suggested Materials:

- Short videos (like "What Is AI?" https://www.youtube.com/watch?v=J4RqCSD--Dg) or pictures that show examples of AI (smartphones/devices, Roombas, robots in factories, etc.).
- Poster board or large paper.
- Markers, crayons, or colored pencils.

Procedure

Introduction (5 minutes):

a. Whole group: Begin by asking students if they know what a robot is or if they've seen any robots in movies, books, or TV shows.
b. Show students pictures or illustrations of robots, smart devices, and computers. Ask what these objects are and what they can do.

Class Discussion (10 minutes):

a. Simplify the concept of AI by saying it is like a smart helper that can think and learn, just like how people learn new things.
b. Show a short video or use images to show students some examples of technology that uses AI.

Activity (15 minutes) Create an AI Helper:

a. Divide the children into groups of three to four and ask them to draw and name their own AI helper or robot friend on the poster board. Encourage creativity and kindness in their designs.
b. Have the groups share their AI helper with the class, explaining its name, what it can do, and how it helps people.

Conclusion (5 minutes):

a. Summarize emphasizing the importance of using AI to help others and being kind to technology.

Assessment:

a. The group activity can be used to assess student learning concerning the lesson objective.

Grade 1

Character Education

TOC Title: G-1 Good Character

Lesson Title: G-1 Good Character Toolbox

Grade Level: 1

Duration: 35–40 minutes

Objective:

- Students will understand the traits of good character.
- Students will be able to demonstrate good character traits in their lives.

Suggested Materials:

- Books featuring characters that demonstrate honesty, kindness, respect, sharing, and/or responsibility, such as Kindness Is My Superpower by Alicia Ortega or Our Class Is a Family by Shannon Olsen, or another book that addresses the topic at an age-appropriate level.

Procedure

Introduction (10 minutes):

a. Start by introducing the concept of a toolbox. Show students a real toolbox or pictures of one. Explain that just like a toolbox holds tools that help us fix things, we have our own "character toolbox" that helps us be good people.
b. Introduce different "tools" as aspects of good character. For example:
   - Honesty tool: when someone is truthful in the things they say and do.
   - Kindness tool: talk about what kindness means and how students can be kind to others.
   - Respect tool: explain the importance of respecting others' feelings and belongings.
   - Sharing tool: emphasize the value of sharing with others.
   - Responsibility tool: discuss why it is important for people to be responsible for their actions.

Read Aloud (10 minutes) [Optional if reading books are not available]:

a. Read a book where a character shows one or more of the toolbox traits (Suggested: Kindness Is My Superpower or Our Class Is a Family).

Class Discussion (5 minutes):

a. Briefly discuss with the students which "tools" they noticed characters using in the story.

Activity (5–10 minutes):

a. Have student volunteers act out scenarios where characters display kindness in various situations. Let students take turns playing different roles to understand the impact of kindness.

  * It may help if the teacher demonstrates this first, e.g., pretend to hold a door open for someone, lend or return something to another student, etc.

Conclusion (5 minutes):

a. Review the different character traits as "tools" and emphasize how these can help them in their daily lives. Ask them to think about which tools they would like to use more often and why.

Homework (Optional):

a. Ask students to talk about the tools of good character with their families.

Assessment:

a. An informal assessment of student participation through observation during class activities and discussions.

Extension Activities:

a. Kindness Challenge: Develop a kindness challenge where students actively seek out ways to be kind to others throughout the week and share their experiences afterward.

Acceptable Behavior

TOC Title: G-1 Acceptable Behavior

Lesson Title: G-1 Acceptable Behavior

Grade Level: 1

Duration: 30–40 minutes

Objective:

- Students will understand and be able to demonstrate acceptable behavior in multiple situations.

Suggested Materials:

- Whiteboard or chart paper.
- Markers.
- Storybooks about behavior such as David Goes to School by David Shannon or another book that addresses acceptable behavior at an age-appropriate level.

Procedure

Introduction (5 minutes):

a. Engage students in a short discussion about what they think 'acceptable behavior' means. List their ideas on the whiteboard or chart paper.

Read Aloud (5–10 minutes) [Optional if reading books are not available]:

a. Read a storybook that illustrates good behavior and discuss it afterward. Encourage students to identify the behaviors exhibited by the characters.

Class Discussion (5–10 minutes):

a. Discuss why rules are important at school. Ask students to share examples of rules they follow in their classroom.
b. Talk about David's behavior in the story. Ask students to identify instances where David broke the rules and what the consequences were.

Activity (10 minutes):

a. Classroom Rules: Together, create a list of classroom rules that promote acceptable behavior. Encourage students to contribute ideas.
b. Write down these rules and display them in the classroom.

Conclusion (5 minutes):

a. Review the lesson by revisiting the classroom rules created by the students.
b. Ask students to reflect on why these rules are important and how they contribute to a positive learning environment.

Assessment:

a. Assessment can be informal, through observation during activities and discussions, taking note of students' participation, understanding, and application of acceptable behavior.

Anti-Bullying

TOC Title: Cyberbullying Lesson Grade 1

Lesson Title: Cyberbullying Lesson Grade 1
Grade Level: 1
Duration: 40–45 minutes

Objective:

- To introduce the concept of cyberbullying.
- Teach empathy and the importance of being kind online.

Suggested Materials:

- Optional – picture books about kindness and friendship.
- Large paper or whiteboard.
- Crayons/markers.
- Internet safety posters.

Procedure

Introduction (10 minutes):

a. Facilitate a class discussion on feelings. Encourage students to share about when they have felt happy, sad, or scared.
b. If reading a picture book, read aloud and discuss with students the characters' feelings throughout the story.

Lesson (10–15 minutes):

a. What Is Cyberbullying? Define cyberbullying. Offer examples students will relate to, e.g., saying mean things online or excluding someone from a game or a chat. (See Chapter 1 content.)
b. Discussion: Encourage students to share any experiences they have had. Emphasize the importance of being kind both in person and online.

Activity (15 minutes):

a. Pass out paper and markers or crayons. Students draw pictures of kind things they can do for others, both online and offline.
b. Have volunteers show their drawings and explain the act of kindness they drew.

Conclusion (5 minutes):

a. On large paper, write out Internet safety rules, i.e., never give out personal information online, always be kind to others, and report unkind/cyberbullying activity to a trusted adult.

TOC Title: G1 Physical Bullying
Lesson Title: Physical Bullying
Grade Level: 1
Duration: 40 minutes

Objectives:

- Students will be able to identify physical bullying behaviors, understand empathy and why it is important when dealing with bullying situations, and develop strategies to prevent and respond to physical bullying.

Suggested Materials:

- Whiteboard and markers.
- Pictures that depict different emotions.
- Large paper or poster board.
- Crayons or markers for posters.

Procedure

Introduction (15 minutes):

a. Begin with a discussion about what makes a good friend and how it feels when someone is unkind or hurts them physically.
b. Explain what physical bullying means. Use examples like hitting, pushing, or taking things by force.
c. Show pictures depicting various emotions. Have students identify these emotions and describe what they feel like. Link them to situations where they might feel that way due to bullying.
d. Talk with the class about what to do if they are bullied or witness someone else being bullied.

Activity (15 minutes) Kindness Posters:

a. Divide the class into small groups (three to four).
b. Pass out large paper or posters and markers/crayons.
c. Have groups create kindness posters that promote friendship, empathy, kindness, and standing up to bullies.

d. Have groups share their posters with the class.
e. Hang the posters around the classroom.

Conclusion (10 minutes):

a. As a group, recap the main points of the lesson. Emphasize the importance of empathy in dealing with physical bullying.

Assessment:

a. Informal assessments can be made based on the groups' posters and individual participation in class discussions and activities.

TOC Title: Grade 1 Verbal Bullying
Lesson Title: Grade 1 Verbal Bullying
Grade Level: 1
Duration: 35–50 minutes

Objective:

- Students will recognize and understand verbal bullying.
- Students will know how to respond appropriately if they experience or witness verbal bullying.

Suggested Materials:

- Suggested picture book for read aloud Llama Llama and the Bully Goat by Anna Dewdney or another book that addresses verbal bullying at an age-appropriate level.
- Paper and markers/crayons for art activity.

Procedure

Introduction (10 minutes):

a. Discuss feelings. Ask students to share how they feel when someone says unkind words to them. Demonstrate empathy and understanding of different emotions as students respond.
b. Define verbal bullying in simple terms suitable for first graders, e.g., When someone repeatedly says mean things about someone else because they want to make them feel bad, use relatable examples they might encounter in daily life.

Read aloud (5–10 minutes):

a. Suggested picture book for read aloud Llama Llama and the Bully Goat by Anna Dewdney or another book that addresses verbal bullying at an age-appropriate level.

Class Discussion (5 minutes):

a. Story Reflection: Discuss the story's characters, their feelings, and the consequences of bullying behavior.
b. Discussion: encourage students to share personal experiences or observations about verbal bullying without pressure to disclose private information.

Activity (10–15 minutes):

a. Pass out art supplies.
b. Individually or in small groups (two to four students), have students create posters or drawings that show ways to stop verbal bullying and promote kindness.
c. Ask students/groups to share their artwork and explain what it means.

Conclusion (5 minutes):

a. Ask students to reflect on what they've learned and how they can apply it in their daily lives.

Assessment:

a. Assess participation during discussions and activities. Focus on the anti-bullying aspect of the lesson and understanding the importance of kindness.

TOC Title: Grade 1 Emotional Bullying

Lesson Title: Understanding and Managing Emotions
Grade Level: 1
Duration: 45 minutes

Objective:

- Students will learn about emotions, empathy, and what to do if they experience emotional bullying.

Suggested Materials:

- Whiteboard or chart paper.
- Markers.
- Pictures depicting various emotions (happy, sad, angry, scared, etc.).

Procedure

Introduction (10 minutes):

a. Have a brief discussion about feelings. Ask students how they feel today.
b. Use pictures to introduce different emotions. Talk about each emotion and ask students to share experiences related to each feeling.

Bullying/Empathy Activity (20 minutes):

a. Divide the class into small groups (three to four students).
b. Have each group choose a picture depicting an emotion.
c. Groups create a drawing or story showing how someone might experience that feeling because of bullying. Encourage students to be creative. Point out that if they know what that emotion/experience feels like, they have empathy for that person/character.
d. Have each group share their picture/story and explain what they think the person/character feels.

Discussion (10 minutes):

a. Facilitate a discussion about the feeling depicted in each picture/story.
   Possible guiding questions: "How did it feel to imagine someone else's feelings?"
   "Why is it important to understand the feelings of others?"
   "What are some things you can do to help someone who is emotionally bullied?"
b. Draw students' attention to the importance of empathy in understanding and helping others.

Conclusion (5 minutes):

a. Highlight the importance of being kind and understanding the feelings of others.
b. Encourage the class to be empathetic by being a good friend and helping people who feel sad or upset.

Homework (Optional):

a. Create a kindness card for someone who may be feeling sad.

Assessment:

a. Informal assessment can be done throughout the lesson by observing students' participation in discussions, their willingness to share their thoughts, and their understanding of the key concepts discussed.

TOC Title: Grade 1 Disability Bullying

Lesson Title: Grade 1 Disability Bullying

Grade Level: 1

Duration: 40–50 minutes

Objective:

- Teach students about disabilities, promote empathy, and prevent bullying based on differences.

Suggested Materials:

- Picture book Just Ask! Be Different, Be Brave, Be You by Sonya Sotomayor or another picture book on disabilities or assistive devices and kindness.
- Images or props that illustrate a range of assistive devices, such as wheelchairs, corrective eyewear, and hearing devices, among others.
- Drawing materials.
- Whiteboard/markers.

Procedure

Introduction (10 minutes):

a. Ask students questions about differences. Examples: "What makes each of us unique?" "How does it feel when someone is unkind to us because of our differences?"

Read Aloud (10 minutes) Suggested:

a. Just Ask: Be Different, Be Brave, Be You by Sonya Sotomayor or another picture book on disabilities or assistive devices and kindness.

Class Discussion (5–10 minutes):

a. Lead a brief discussion on acceptance and being kind to others regardless of their differences. Ask students to share how they get to know other people (talking to them, playing with them, etc.).
b. Guide students, if needed, to the conclusion that becoming friends with people with disabilities is the same as becoming friends with anyone. Be friendly and kind.

Activity (10–15 minutes):

a. Role-playing: Create scenarios where someone with a disability might need help or support. Let students act out the parts of the person in need and the helper, then reverse roles.
b. Discussion: Ask students to share how it felt to be in need, how it felt when someone helped them, and how it felt to be the helper.

Conclusion (5 minutes):

a. Summarize what students learned about disabilities and kindness.
b. Optional: Consider a class pledge where students pledge to be kind to everyone, regardless of differences.

Assessment:

a. Informal assessment of class participation and role-playing activity.

Using Good Judgment:

TOC Title: Grade 1 Good Judgment

Lesson Title: First Grade Good Judgment
Grade Level: 1
Duration: 30–35 minutes

Objective:

- The students will understand the concept of good judgment.
- The students will be able to make wise decisions.

Suggested Materials:

- Technology to show video.
- The Berenstain Bears and the Truth (https://www.youtube.com/watch?v=FNKEo7_lHj8).
- If not doing the video presentation, read aloud a book that shows characters making decisions and using good judgment.
- Whiteboard and markers.
- Paper and crayons.
- Large paper or posters.
- Assessment sheets with one to two simple questions related to the objective (a sample sheet is provided after the lesson plan).

Procedure

Introduction (10 minutes):

a. Class Discussion: What does "judgment" mean? Use first grade-friendly examples, such as choosing between different toys or deciding what game to play.

Video (Or Read Picture Book) (10–15 minutes):

a. After the video or story, discuss the importance of telling the truth and making good choices.

Activity (5 minutes):

a. Decision-Making Tree: on poster board, large paper, or whiteboard create a large decision-making tree. Use simple illustrations or words to represent choices we might make and their consequences (positive or negative).

b. Discuss the different scenarios.

c. Ask students to map out the decisions and their outcomes on the tree.

Conclusion (5 minutes):

a. Review the main points about good judgment and decision-making.

b. Encourage students to apply what they've learned in their daily lives, making good choices and thinking about consequences.

Assessment:

a. Distribute assessment sheets with simple questions related to the lesson (see below).

Good Judgment Scenarios

a. Scenario: Emma finds a wallet with money in it on the playground.
   - Question: What do you think Emma should do with the wallet and money? Why?

b. Scenario: Jacob sees a younger student being teased by other children.
   - Question: What should Jacob do?

Artificial Intelligence:

TOC Title: G-1 Artificial Intelligence Lesson

Lesson Title: G-1 Artificial Intelligence Lesson

Grade Level: 1

Duration: 35 minutes

Objective:

- Students will understand the concept of AI and how we can use it responsibly.

Suggested Materials:

- Pictures of robots and AI devices.
- Whiteboard or chart paper.
- Markers.
- STEM-building toys like Lego, Tinker Toys, blocks, etc. (if available).

Procedure

Introduction (10 minutes):

a. Whole group discussion on robots and AI. Show the class pictures of robots or smart devices. Ask the students what they already know about these things.
b. Explain to the class that it is like a very smart robot that can help people with tasks, and that it is important for us to learn how to use it responsibly.

Activity (15 minutes):

a. Divide the class into small groups (three to four).
b. Allow each group to select paper and markers or building supplies.
c. Ask groups to draw or build and name their friendly robot or AI helper.

d. Encourage discussion about what their robot can do and how it helps people.

Conclusion (10 minutes):

a. Bring the class back together and invite each group to share their robot/AI designs.
b. Lead the students in a discussion about the responsibilities that might come with using AI. Suggested questions: How can we treat robots with kindness? What do you think are some things that robots or AI can learn to do that will help people? How can we make sure AI doesn't hurt anyone's feelings? Encourage the students to think about ways they can be good friends with their robot helpers.

Assessment: Robot Rules:

a. Provide each student with a worksheet or piece of paper divided into sections.
b. Ask them to create their "Robot Friend Rules" by drawing and writing simple statements about how they would responsibly use their AI friend.
c. Offer prompts or sentence starters to guide them, such as:
    - "My robot friend will help by …"
    - "I will be kind to my robot friend by …"
    - "I will make sure my robot friend doesn't …"
    - Encourage creativity and thoughtful responses.

Grade 2

Character Education

TOC Title: G-2 Good Character

Lesson Title: G-2 Good Character

Grade Level: 2

Duration: 50 minutes

Objective:

- The student will understand and be able to identify the traits ("tools") of good character.

Suggested Materials:

- Whiteboard or chart paper
- Markers
- Pictures or drawings of different tools
- Construction paper
- Scissors and glue
- A toolbox, or picture of a toolbox

Procedure

Introduction (10 minutes):

a. Begin by discussing what a toolbox is (it holds different tools needed for specific tasks).
b. Relate this concept to traits of good character. Explain that just like a toolbox holds tools for different jobs, our character has different traits for various situations.

Class Discussion (15 minutes):

a. Show pictures or drawings of different tools (hammer, screwdriver, tape measure, etc.). Discuss each tool's purpose and how it helps with specific tasks.
   - Relate each tool to a specific character trait, e.g., hammer for honesty, screwdriver for kindness, tape measure for respect, wrench for responsibility, pliers for compassion, and level for fairness (match the tool and traits in a way

that makes sense to the class). Write the traits and their corresponding tools on the board.

- Engage the students in a discussion about how these character traits apply to everyday situations. Use examples like helping a friend, telling the truth, standing up against bullying, etc.

Activity (15 minutes):

a. Pass out art supplies (construction paper, markers, scissors, glue).
b. Have students draw and cut out tools (or use pre-made pictures) that represent different character traits discussed earlier.
c. Have them paste these tools onto a large piece of construction paper, creating their own "Character Toolbox."

Conclusion (10 minutes):

a. Group Discussion: Bring class together with their "toolboxes". Offer different scenarios (e.g., you see the new student sitting alone at lunch). Then ask students to select a tool that would help in that situation (e.g., use their kindness tool or compassion tool – go and sit with them, make them feel welcome).
b. Ask students to share why they selected a particular tool.

Assessment:

a. Evaluate students' understanding through participation in class discussion, and their ability to identify character traits in scenarios. The teacher might also consider evaluating the student creativity exhibited in the character toolboxes.

Extension Activities:

a. Encourage students to find examples of good character traits in stories they read, movies they watch, and in real life and share them with the class.

Acceptable Behavior

TOC Title: G-2 Acceptable Behavior

Lesson Title: G-2 Acceptable Behavior

Grade Level: 2

Duration: 40 minutes

Objective:

- Students will understand what acceptable behavior looks like in various situations and environments.

Suggested Materials:

- Whiteboard and markers.
- Chart paper.
- Markers/crayons.
- Scenario cards. Written situations where behavior needs to be determined as acceptable or not, e.g., listening politely, waiting your turn, sharing, talking out of turn, cutting in line, being physically aggressive toward others.
    - Enough scenarios for pairs or small groups to each have their own scenario.
- Stickers or small rewards for participation (optional).

Procedure

Introduction (10 minutes):

a. Begin by asking the students what they think "acceptable behavior" means. Write their responses on the whiteboard.

Class Discussion (5 minutes):

a. Lead a class discussion about different places (in class, on the playground, at home, etc.). Ask students to identify behaviors that are considered acceptable in these places.
b. Together, create a class agreement about what acceptable behavior is in the classroom. Write down their suggestions on chart paper.

Activity (15 minutes):

a. Divide the class into pairs or small groups (three to four maximum).
b. Hand out scenario cards to small groups or pairs.
c. Each group reads their scenario and discusses whether the behavior in that situation is acceptable. They can act out the scenarios to showcase both acceptable and unacceptable behaviors.
d. Bring the class back together and have each group share their scenarios and conclusions. Discuss why certain behaviors were acceptable or not. (Hand out stickers/rewards for effort and participation during discussion).

Conclusion (10 minutes):

a. Review what was learned about acceptable behavior and the importance of demonstrating it in different settings.
b. Remind students of the class agreement and encourage them to practice acceptable behavior everywhere, not just at school.

Homework (Optional):

a. Ask students to discuss acceptable behavior with their families and share their classroom agreement with them.

Assessment:

a. Observe participation during discussions and understanding shown in scenario activities.

Anti-Bullying

TOC Title: Cyberbullying Lesson Plan Grade 2

Lesson Title: Cyberbullying Lesson Plan Grade 2
Grade Level: 2
Duration: 40 minutes

Objective:

- Introduce the concept of cyberbullying, identify its forms, and empower students with strategies to respond to and prevent cyberbullying.

Suggested Materials:

- Poster board and markers.
- Worksheets or coloring pages.
- Whiteboard and markers.
- Simple, age-appropriate list of Internet safety rules.
- Storybooks or videos about kindness and online behavior (optional).

Procedure

Introduction (10 minutes):

a. Discuss being kind to others. Encourage students to share what it means and how it feels to be kind or unkind to others.
b. Define cyberbullying in age-appropriate terms, i.e., when someone uses unkind or mean words or behavior using computers, phones, or tablets.
c. Optional – read a story or show a video depicting kind, appropriate behavior online.

Class Discussion (10 minutes):

a. Brainstorm together what might be considered cyberbullying, e.g., sending mean messages, spreading rumors online, excluding someone from online games, etc.

b. Give the class realistic scenarios to help them understand what cyberbullying looks like in everyday situations.

Suggested examples:

- You are playing a game online, and one of the other players is saying unkind things to you or calling you mean names.
- You want to play a game online, but the other players are excluding you on purpose.
- Someone has a picture of you that you do not like, and they post it online to make fun of you?
- You keep getting messages from someone you know who makes fun of you or things you like.

c. Ask students to talk about how they would feel if someone said mean things to them online. What would they do if that happened?

Activity (15 minutes):

a. In small groups, have the class create posters that show how to be kind online and what to do if someone experiences or witnesses cyberbullying.

Conclusion (5 minutes):

a. Review cyberbullying and being kind online.
b. Encourage students to talk to their families about what they learned.

Extension:

a. Revisit this concept frequently throughout the year.
b. Include kindness and cyberbullying in-class activities and lessons whenever appropriate.
c. Encourage students to talk about their online experiences and remind them to report any concerning behavior to trusted adults.

TOC Title: Physical Bullying Grade 2

Lesson Title: Physical Bullying Grade 2

Grade Level: 2

Duration: 40 minutes

Objectives:

- Students will be able to define physical bullying.
- Students will know what to do when they witness physical bullying or are bullied themselves.
- Students will understand empathy and why it is important to have empathy for victims of bullying.

Suggested Materials:

- Technology to show a video to the class.
- Large paper or poster board.
- Markers or crayons.

Procedure

Introduction (10 minutes):

a. Discuss feelings and emotions with the class. Show pictures of various emotions (happy, sad, angry, scared, etc.) and ask students to identify and describe these emotions.
b. Guide a conversation about how different situations can make people feel certain emotions.

Video (5 minutes):

a. Suggested video: The Meanest Girl in Second Grade (https://youtu.be/QFWfFCmjH_s?si=zmZ0JT-vXY5OWRqA) or another short age-appropriate video on the topic of physical bullying.

Class Discussion (5 minutes):

a. Lead the class in discussing what they noticed in the video using discussion questions such as:
   - What was Zoe like? How do you think the other children felt around Zoe?

- Why did they start to ignore her?
- Can you think of a time when you felt like one of the characters in the video? If so, talk about it.

Activity (15 minutes):

a. Individually or in small groups, students will create a poster.
b. Provide paper and art supplies to the students.
c. Ask them to draw a picture showing a bullying situation and a possible solution that will help the victim feel better and help the bully realize why that behavior is wrong and how it affects others.
d. Encourage them to include a short caption or sentence explaining their drawing.

Conclusion (5 minutes):

a. Summarize the key points of the lesson: what physical bullying is and why empathy is important.
b. Encourage students to practice empathy by being kind and understanding toward others (even bullies).

Homework (Optional):

a. Provide a simple worksheet or journal prompt asking students to write or draw about a time they showed empathy toward someone else or how they would handle a bullying situation.

Assessment:

a. Review the drawings created by the students and ask volunteers to explain their artwork, emphasizing how their drawing relates to empathy and kindness.

TOC Title: Grade 2 Verbal Bullying

Lesson Title: Grade 2 Verbal Bullying

Grade Level: 2

Duration: 45–55 minutes (split into two lessons if needed. Stop after the read aloud, then discuss the story, and do a drawing activity in the second session).

Objectives:

- Students will understand the effects of verbal bullying.
- Students will be able to identify strategies to handle and prevent verbal bullying.

Suggested Materials:

- Picture book where the main character shows strong anti-verbal bullying strategies, such as Stand Tall Molly Lou Melon by Patty Lovell or another book that addresses the topic at an age-appropriate level.
- Drawing Materials: Large paper, markers/crayons.

Procedure

Introduction (10 minutes):

a. Start by talking about feelings. Ask students how certain words or actions make them feel.
b. Bring up bullying, explaining that sometimes words can hurt just as much as actions.

Read Aloud (10–15 minutes) [Optional if reading books are not available]

a. Stand Tall Molly Lou Melon by Patty Lovell, or another picture book with anti-verbal bullying strategies.

Class Discussion (5–10 minutes):

a. Lead a discussion about the story's message. Ask questions like:
    - What unkind things did the bully or bullies say to the main character?

- How did the characters respond to their bullying?
- Where do you think the main character learned to handle bullies?

Activity (15 minutes):

a. Drawing Activity: Provide drawing materials. Ask students to draw a scene from the book where the main character stands up to the bully or bullies, or a scene where they've faced bullying themselves.

Conclusion (5 minutes):

a. Ask students to reflect on what they learned today about verbal bullying and standing up for themselves.

Assessment:

a. Evaluate students' participation in discussions and how their drawing activity connects with the message of the lesson.

TOC Title: Emotional Bullying Grade 2

Lesson Title: Emotional Bullying
Grade Level: 2
Duration: 40 minutes

Objectives:

- Define emotional bullying.
- Identify examples of emotional bullying.
- Cultivate empathy toward others who might be experiencing emotional bullying.

Suggested Materials:

- Whiteboard or chart paper
- Markers
- Drawing materials (paper, crayons, etc.)
- Story book illustrating emotions and empathy (Suggestion: The Juice Box Bully by Bob Sornson and Maria Dismondy)

Procedure

Introduction (10 minutes)

a. Begin with a class discussion on feelings/emotions. Ask students how they feel when someone says something mean to them.
b. Define emotional bullying emphasizing that it involves hurting someone's feelings on purpose.
c. List examples of emotional bullying on the board (e.g., name-calling, excluding someone on purpose, and spreading rumors).

Read Aloud (10 minutes) [Optional if reading books are not available]

a. Read a story that depicts emotional bullying and how it affects the target/victim.

Class Discussion (5 minutes)

a. Open a discussion about how showing empathy can help prevent emotional bullying. Encourage students to share personal experiences (if they're comfortable) and discuss how they would feel if they were in the bullied person's shoes.

Empathy Activity (10 minutes):

a. Divide the class into groups of 3–4.
b. Provide scenarios of emotional bullying or hurtful situations for them to discuss.
c. Ask each group to come up with ways they could help the person being bullied to feel better. Encourage empathy and understanding of others' feelings.
d. Have each group share their ideas with the class.

Conclusion (5 minutes):

a. Summarize emotional bullying and empathy.
b. Encourage students to practice empathy and kindness toward others and remind them to seek help from adults if they witness or experience emotional bullying.

Assessment:

a. Distribute a worksheet or activity where students can identify examples of emotional bullying. Have students suggest appropriate responses to help the person being bullied.
b. Review the worksheet together and discuss the answers as a class to reinforce understanding.

TOC Title: Grade 2 Disability Bullying Lesson

Lesson Title: Grade 2 Disability Lesson

Grade Level: 2

Duration: 45–55 minutes

Objective:

- The students will gain an understanding of disabilities.
- The students will recognize the importance of and be able to promote empathy and prevent bullying based on differences.

Suggested Materials:

- Picture books featuring characters with disabilities (e.g., Emmanuel's Dream: The True Story of Emmanuel Ofosu Yeboah by Laurie Ann Thompson and Sean Qualls, or My Brother Charlie by Holly Robinson Peete and Ryan Elizabeth Peete, or another book that addresses the topic at an age-appropriate level).
- Images or drawings depicting various assistive devices (e.g., wheelchairs, crutches, hearing aids, passive noise isolation (PNI) headphones).
- Paper and pencils for a writing/drawing activity.
- Role-play scenarios related to disability bullying.

Procedure

Introduction (10 minutes):

a. Discuss people's differences. Ask the students to share what they know about disabilities.
b. Read a picture book that includes a character with a disability and assistive devices, prompting discussions about the character's experiences and feelings.
c. Introduce various disabilities through images or drawings, explaining each briefly and highlighting that everyone is unique.

Read Aloud (10–15 minutes):

a. Read one of the suggested picture books or another appropriate story that has a main character with a disability.

Class Discussion (10 minutes):

a. Lead a discussion focusing on key points:
   - Ask students to identify the challenges the main character(s) faced and how they overcame them.
   - Discuss the role of determination and resilience in the main character's life. What kept them going?
   - Talk about the importance of perseverance and resilience.
   - If appropriate, talk about the impact of the character's actions on their community and beyond.

Activity (10–15 minutes):

a. Have students reflect on a challenge each of them has faced and how they handled it. Then ask them to write a short paragraph or draw a picture representing their challenge and how they overcame it.
b. Encourage students to share their reflections if they are comfortable doing so to foster empathy and understanding between the students.

Conclusion (5 minutes):

a. Summarize the lesson by emphasizing the importance of kindness, empathy, and inclusion.
b. Discuss strategies such as talking to a teacher/trusted adult or showing support if they witness or experience disability bullying.

Assessment:

a. An informal assessment of student learning can be made using the students' reflective writing or pictures.

Using Good Judgment:

TOC Title: Grade 2 Good Judgment
Lesson Title: Grade 2 Good Judgment
Grade Level: 2
Duration: 40–45 minutes

Objective:

- Students will understand good judgment and be able to apply it in various situations.

Suggested Materials:

- Copy of The Three Questions by Jon J. Muth or another picture book that depicts characters making decisions and using good judgment.
- Scenarios that require decision-making are written on large index cards.

Procedure

Introduction (10 minutes):

a. Ask students what judgment is and why it is important.
b. Explain that good judgment means making wise decisions and choices that consider consequences and others' feelings.
c. Describe simple scenarios where good judgment is needed (e.g., sharing toys, resolving conflicts, etc.).

Read Aloud (10–15 minutes):

a. Read a story to the class.
b. Stop at key points to discuss characters' decisions and ask questions:
   "What would you do in this situation?"
   "How could the character show better judgment?"
   "What might happen if they made a different choice?"

Activity (15 minutes):

a. Divide students into pairs or small groups.
b. Provide scenarios on index cards where good judgment is needed.
c. Encourage students to act out the scenarios, make decisions, and discuss their choices afterward.

Conclusion (5 minutes):

a. Recap the main points about good judgment.
b. Encourage students to apply what they have learned in their daily lives.

Assessment:

a. Make an informal assessment of student learning during decision-making scenarios, activities and discussions during the read aloud.

Artificial Intelligence

TOC Title: G-2 Ethics and Artificial Intelligence

Lesson Title: G-2 Artificial Intelligence Ethics

Grade Level: 2

Duration: 40 minutes

Objective:

- Students will understand the concept of AI and be able to explain the importance of using AI ethically and responsibly.

Suggested Materials:

- Whiteboard or flipchart.
- Markers.
- Pictures or illustrations of AI-related technology (robots, smart devices, etc.).
- Storybooks or videos about AI (optional).

Procedure

Introduction (10 minutes):

a. Whole Group: Ask students if they have heard of smart devices like Alexa or Siri.
b. Show pictures or illustrations of different AI-related technologies and ask students to identify or describe what they see.
c. Explain that these are examples of AI, which stands for artificial intelligence.

Class Discussion (5 minutes):

a. Explain that AI helps machines or computers learn and make decisions as humans do.
b. Use examples students will relate to, e.g., how some games or apps learn from how we use them to become better.
c. Encourage questions from the students about AI.

Activity (15 minutes):

a. Ask students to draw a picture and/or write a few sentences about a friendly robot or AI helping someone.
b. Encourage them to think about how this AI would act fairly and kindly.
c. Circulate as students work and ask them questions about their drawings or stories.

Conclusion (10 minutes):

a. Invite some students to share their drawings or stories with the class.
b. Recap the importance of using AI fairly and kindly.

Homework (Optional):

a. Encourage students to share what they've learned about AI with their friends and family.

Assessment:

a. An informal assessment of student understanding can be done during the activity as students draw, write, and talk about their robot friends.

Notes

1 Betkowski, A., August 01, 2023, “Teaching Tuesday: What Is Character Education?,” Teaching & School Administration, Grand Canyon University, www.gcu.edu/blog/teaching-school-administration/what-character-education#:~:text=Character%20education%20aims%20to%20develop,a%20world%20worth%20living%20in, retrieved November 12, 2023.

2 OpenAI, August 2023, “How Do We Define Character Education?” GPT-3.5, https://chat.openai.com/c/e15869b4-2055-44f6-a69e-76aea0a13b40, queried November 12, 2023.

3 American Institutes for Research, 2023, “Essential Components of MTSS,” American Institutes for Research, https://mtss4success.org/essential-components#:~:text=A%20multi%2Dtiered%20system%20of,from%20a%20strengths%2Dbased%20perspective, retrieved November 10, 2023.

4 Center on PBIS, 2023, “What Is PBIS?” Center on PBIS, Positive Behavioral Interventions & Supports, www.pbis.org, www.pbis.org/pbis/what-is-pbis, retrieved November 10, 2023.

5 Interval Technology Partners, LLC, 2021, “RTI vs. MTSS,” Interval Technology Partners, LLC, www.enrichingstudents.com/rti-vs-mtss/#:~:text=RTI%20is%20considered%20a%20more,%2C%20and%20social-emotional%20support, retrieved November 10, 2023.

6 Character.org, (n.d.), “11 Principles in Schools,” Character.org (formerly the Character Education Partnership), https://character.org/11-principles-in-schools/, retrieved November 10, 2023.

7 “Toolbox with Many Tools,” <image>: “Freepik.com”. Designed by brgfx. This image has been designed using assets from Freepik.com, Freepik Company S.L., www.freepik.com, used with permission, and in compliance with the license agreement: https://www.freepik.com/free-vector/toolbox-with-many-tools_6027743.htm#query=kids%20toolbox&position=49&from_view=keyword&track=ais (Image by brgfx).

8 Office of the Deputy Secretary (ED), March 2002, “U.S. Department of Education Strategic Plan, 2002–2007,” Office of the Deputy Secretary (ED), Washington, DC, pp. 6 and 16, https://www.govinfo.gov/content/pkg/ERIC-ED466025/pdf/ERIC-ED466025.pdf, retrieved November 12, 2023.

9 U.S. Department of Education, (n.d.), “Character Education…Our Shared Responsibility,” U.S. Department of Education, Office of Safe and Drug-Free Schools, www2.ed.gov/admins/lead/character/brochure.html, retrieved November 12, 2023.

10 Sugiarti, R., Erlangga, E., Suhariadi, F., Winta, M. V. I., & Pribadi, A. S., April 27, 2002, “The Influence of Parenting on Building Character in Adolescents,” *Heliyon*, 8(5), e09349. https://doi.org/10.1016/j.heliyon.2022.e09349. PMID: 35586332; PMCID: PMC9108886, open access article under the CC BY-NC-ND license (http://creativecommons.org/licenses/by-nc-nd/4.0/), retrieved November 12, 2023.

11 The Foundation of Character-Core Values, The graphic in Figure 1.2 was designed and produced by the authors.

12 The 14 core values were compiled and summarized from multiple sources, including (1) Carpenter, M., March 11, 2022, "What Are Core Values, and How Do You Pick Them for Your Characters?," https://medium.com/keyboard-quill/what-are-core-values-and-how-do-you-pick-them-for-your-characters-aa3b85cf3b68#:~:text=A%20core%20value%20is%20a,guided%20by%20their%20core%20values, (2) U.S. Department of Education, Office of Safe and Drug-Free Schools, (n.d.), "Character Education…Our Shared Responsibility,", www2.ed.gov/admins/lead/character/brochure.html, (3) The National Association of Intercollegiate Athletics (NAIA), (n.d.), "Five Core Values," www.naia.org/champions-of-character/five-core-values, (4). Character Counts!, (n.d.), "The Six Pillars of Character," https://charactercounts.org/six-pillars-of-character, (5) OpenAI, 2024 at www.openai.com, (6) Spallino, J., January 23, 2017, "How Character Education Helps Kids Learn and Develop," Service Learning, www.methodschools.org/blog/how-character-education-helps-kids-learn-and-develop, and (7) Sugiarti, R., Erlangga, E., Suhariadi, F., Winta, M. V. I., Pribadi, A. S., April 27, 2002, "The Influence of Parenting on Building Character in Adolescents," *Heliyon*, 8(5), e09349. https://doi.org/10.1016/j.heliyon.2022.e09349. PMID: 35586332; PMCID: PMC9108886.

13 Bicard, S. C, Bicard, D. F., & the IRIS Center, 2012, "Defining Behavior," http://iris.peabody.vanderbilt.edu/case_studies/ICS-015.pdf, retrieved on November 13, 2023.

14 Kid Sense, (n.d.), "What Is behaviour?" Kid Sense, https://childdevelopment.com.au/areas-of-concern/behaviour/#:~:text=Behaviour%20refers%20to%20how%20one,to%20everyday%20environments%20and%20situations, retrieved November 13, 2023.

15 UNICEF, November 2021, "Defining Social Norms and Related Concepts," UNICEF Child Protection Programme Team and Social and Behavior Change Team, https://www.unicef.org/media/111061/file/Social-norms-definitions-2021.pdf, retrieved November 13, 2023.

16 Acceptable K-2 Student Behaviors were compiled and summarized from multiple sources including: (1) Clarence High School, (n.d.), "Positive Behavior Expectations for Elementary Students, www.clarenceschools.org/Page/1912#:~:text=Follow%20class%20directions%20and%20expectations,Accept%20positive%20and%20negative%20consequences, retrieved November 13, 2023; (2) Newfoundland and Labrador, Canadian Province, "Behavioural Expectation Samples, 2009, Safe and Caring Schools Resource Guide, www.gov.nl.ca/education/files/k12_safeandcaring_teachers_pbs_behaviouralexpectations.pdf, retrieved November 13, 2023; and (3) Center for Teaching Innovation, (n.d.), "Getting Started with Establishing Ground Rules,", Cornell University, https://teaching.cornell.edu/resource/getting-started-establishing-ground-rules, retrieved November 14, 2023.

17 Unacceptable K-2 Student Behaviors were compiled and summarized from multiple sources, including: (1) University of Adelaide, (n.d.), "Examples of Inappropriate Behaviour," www.adelaide.edu.au/student/behaviour/examples-of-inappropriate-behaviour#:~:text=Behaviours%20that%20are%20considered%20to,or%20have%20mental%20health%20issues, retrieved November 14, 2023; (2) Pathway2success, March 3, 2020, "Managing Disrespectful & Rude Behaviors in the Classroom," www.thepathway2success.com/managing-disrespectful-rude-behaviors-in-the-classroom/#:~:text=The%20child%20who%20talks%20back,of%20laughter%20in%20the%20room, retrieved November 14, 2023; and (3) University of Cambridge, (n.d.), "Acceptable and Unacceptable Behaviour," www.hr.admin.cam.ac.uk/policies-procedures/dignity-work-policy/guidance-managers-and-staff/guidance-managers/acceptable-and#:~:text=Unacceptable%20behaviour%20(including%20bullying%2C%20harassment,or%20involve%20groups%20of%20people, retrieved November 14, 2023.

18 Riverside Medical Clinic, (n.d.), "Statistics and Laws," www.rmccharity.org/bullying-prevention-institute/resources/facts-and-laws, retrieved December 2, 2023.

19 Bischoff, P., March 25, 2022, "Almost 60 Percent of Parents with Children Aged 14 to 18 Reported Them Being Bullied," Comparitech, www.comparitech.com/blog/vpn-privacy/boundless-bullies, retrieved December 2, 2023.

20 Patchin, J., & Hinduja, S., 2020, "Tween Cyberbullying in 2020," Cyberbullying Research Center, chrome- https://i.cartoonnetwork.com/stop-bullying/pdfs/CN_Stop_Bullying_Cyber_Bullying_Report_9.30.20.pdf.

21 Patchin, J., October 4, 2023, "Cyberbullying Continues to Rise among Youth in the United States," https://cyberbullying.org/cyberbullying-continues-to-rise-among-youth-in-the-united-states-2023#:~:text=Somewhat%20surprisingly%2C%20the%20percentage%20of,%25%20and%2025%25%20respectively), retrieved December 2, 2023.

22 The Anti-Bullying Alliance, (n.d.), "The Anti-Bullying Alliance," https://anti-bullyingalliance.org.uk/aba-our-work, retrieved October 20, 2023.

23 The Centers for Disease Control and Prevention, (n.d.), "Fast Facts: Preventing Bullying,", (n.a.), https://www.cdc.gov/youth-violence/about/about-bullying.html?CDC_AAref_Val=https://www.cdc.gov/violenceprevention/youthviolence/bullyingresearch/fastfact.html retrieved March 12, 2024.

24 (n.a.) (n.d.) StopBullying.gov, www.stopbullying.gov/resources/laws, retrieved October 20, 2023.

25 (n.a.) January 25, 2023, University of the People, (n.d.), "Definition of Bullying," www.uopeople.edu/blog/definition-of-bullying/, retrieved October 20, 2023.

26 (n.a.) (n.d.) Preventing and Promoting Relationships & Eliminating Violence Network, (n.d.), "Types of Bullying," www.prevnet.ca/bullying/types, retrieved November 1, 2023.
27 The Centers for Disease Control and Prevention, (n.d.), "Fast Facts: Preventing Bullying," https://www.cdc.gov/violenceprevention/youthviolence/bullyingresearch/fastfact.html#:~:text=Bullying%20is%20a%20frequent%20discipline,and%20primary%20schools%20(9%25), retrieved November 27, 2023.
28 UNESCO, (n.d.), www.unesco.org/en/days/against-school-violence-and-bullying, retrieved November 2, 2023.
29 The Centers for Disease Control and Prevention, (n.d.), "Fast Facts: Preventing Bullying," www.cdc.gov/violenceprevention/youthviolence/bullyingresearch/fastfact.html#:~:text=Bullying%20is%20a%20frequent%20discipline,and%20primary%20schools%20(9%25), retrieved November 27, 2023.
30 Lee, A. M. I., "The Difference between Teasing and Bullying," www.understood.org/en/articles/difference-between-teasing-and-bullying, retrieved November 24, 2023.
31 Johnson, N., Stixrud, W., & Psychology Today, August 18, 2022, "Helping Kids Become Good Decision Makers," www.psychologytoday.com/us/blog/the-self-driven-child/202208/helping-kids-become-good-decision-makers, retrieved November 13, 2023.
32 Neuroscience News, (n.d.), "Child's Play: Kids as Young as Six Consider Choices in Moral Judgments," https://neurosciencenews.com/moral-judgment-child-23350/, retrieved November 27, 2023.
33 Marcella, A., "Child Flipping a Coin into the Air," ChatGPT-4, developed by OpenAI, generated and retrieved January 19, 2024.
34 Wellspring Center for Prevention , (n.d.), "Tips for Helping Children Develop Healthy Decision-Making Habits," , https://wellspringprevention.org/blog/help-child-develop-decision-making-skills, retrieved November 29, 2023.
35 Neuroscience News, (n.d.), "Child's Play: Kids as Young as Six Consider Choices in Moral Judgments," https://neurosciencenews.com/moral-judgment-child-23350, retrieved November 27, 2023.
36 Blackboard Photo by Jason Dent on Unsplash at https://unsplash.com/@jdent?utm_content=creditCopyText&utm_medium=referral&utm_source=unsplash (Jason Dent) on https://unsplash.com/photos/white-and-black-abstract-painting-itA0ybdIDTs?utm_content=creditCopyText&utm_medium=referral&utm_source=unsplash (Unsplash). Original photo was redesigned, enhanced and text added by author, retrieved on November 16, 2023.
37 NIST, January 21, 2020, "Trustworthy AI: A Q&A with NIST's Chuck Romine," www.nist.gov/blogs/taking-measure/trustworthy-ai-qa-nists-chuck-romine, retrieved November 16, 2023.

38 Cotton, P., Patel, M., & Wei, W., May 2022, "The Foundational Standards for AI ISO/IEC 22989 and ISO/IEC 23053," ISO/IEC AI Workshop, https://jtc1info.org/wp-content/uploads/2022/06/03_08_Paul_Milan_Wei_The-foundational-standards-for-AI-20220525-ww-mp.pdf, retrieved November 16, 2023.
39 Marcella, A., November 16, 2023, "Please Explain What Generative AI Is Using Non-Technical Terms," ChatGPT 3.5, OpenAI, https://chat.openai.com/share/cc8db088-0c4f-4426-a3d2-b4763ff82e12, retrieved November 16, 2023.
40 Kittelstad, K., (n.d.), "What's the Difference between Ethics, Morals and Values?" https://examples.yourdictionary.com/difference-between-ethics-morals-and-values.html, retrieved November 18, 2023.
41 OpenAI, (n.d.), "Provide a Brief Definition of Norms as Used in the Field and Study of Ethics," GPT-3.5, OpenAI's Large-Scale Language-Generation Model, https://chat.openai.com/c/6c151798-9d2d-482b-a075-db136c385112, queried November 18, 2023.
42 The Core Principles of Ethics, Figure 1.7, was designed and produced by the authors.
43 Velasquez, M., Moberg, D., Meyer, M., Shanks, T., et al., November 5, 2021, "A Framework for Thinking Ethically," The Markkula Center for Applied Ethics at Santa Clara University, www.scu.edu/ethics/practicing/decision/framework.html.
44 Barry, P., October 12, 2023, "The Meaning Behind the Song: Teach Your Children by Crosby – Stills – Nash & Young," https://oldtimemusic.com/the-meaning-behind-the-song-teach-your-children-by-crosby-stills-nash-young, retrieved November 21, 2023.

2
Nurturing Digital Citizens
Cyber Safety for Early Learners

Introduction

As we discussed in Chapter 1, technology brings many benefits to users who have learned to use technology responsibly. Awareness, care, caution, and prudent use are the guiding principles when approaching and using technology. Also discussed is the reality that technology (in its many forms) also presents potential risks to users when used carelessly or without taking the proper precautions and safeguards. Users who fail to act responsibly or use caution when embracing technology, whether for business, use in the classroom, or simply for recreational activities, may find themselves open to unwanted and oftentimes unseen risks.

This chapter presents a broad examination of the cyber risks that users face when engaging with society's increasing dependence on technology. Here we use the very broad term "users," which we intentionally and generically define as adults and children.

The primary focus of this chapter is on fundamental cyber safety concepts, risks specific to early learners, and practical recommendations for educators working with kindergarten through second-grade students. The emphasis is on age-appropriate strategies, collaboration with parents, and real-life examples to enhance the applicability and impact of cyber safety education in early childhood education settings.

The objective is to provide students with the knowledge and confidence to embrace technology, understand technology, and use technology, while doing so responsibly, safely, and prudently.

Examples via class lesson plans and exercises are provided to assist the educator in discussing the broader concept of cyber safety with students.

DOI: 10.1201/9781003465928-2

The Digital Landscape for Early Learners

The Pervasiveness of Technology in Education

> Digital technology is transforming the world of work. To produce the knowledge workers of tomorrow, and to maximize the ability of children to learn, it must also be allowed to transform the world of education.[1]

Technology has become deeply integrated across all aspects of modern education, transforming how instruction is delivered and how students learn. Seamless blending of physical and virtual interactions allows learning to transcend barriers and creates digitally augmented, blended instruction models. This represents a fundamental shift as extensive technological immersion makes education the prime territory for deploying pervasive computing.

Easy access to vast information and data-driven insights certainly confers valuable advantages, as do efficiencies in communication and administrative tasks. However, uncontrolled overuse risks jeopardizing accrued intergenerational wisdom and shared humanity. Keeping developmental needs and learning goals at the center can help overcome such pitfalls. Moderation and balance will thus remain key guiding principles moving forward.

Ubiquitous access to devices, platforms, and tools enhances interactivity, expands access to information, and enables customized learning experiences catering to diverse needs. Students today have opportunities to learn collaboratively across geographies, receive personalized guidance, and develop critical, future-ready skills. Technology facilitates student-centric active learning and develops essential 21st-century competencies like digital literacy.

Integrating technology throughout instruction equips students with relevant capabilities and prepares them for technology-driven careers. It streamlines administrative processes and provides professional development opportunities for educators as well. Technology-enabled, individual-centric customizations are vital for relevance and success in education.

However, while promising, there are also emerging concerns surrounding equitable access, ethical use, and respecting human connections in learning. Addressing these responsibly while harnessing technology's potential requires mindful, balanced application focused on pedagogical objectives over capabilities alone. Ongoing teacher training and investments to bridge digital divides are equally necessary.

Still, suitably incorporated, educational technology can enrich learning and make it enjoyable, engaging, and effective. By interweaving physical and virtual interactions, learning can transcend spatial, chronological, and social barriers. With sound strategies on fronts like teacher readiness and equitable access, integrated ed tech promises to make quality, personalized learning truly scalable and sustainable.

The path forward warrants continued efforts to enable disadvantaged groups through access mechanisms, alongside policy measures around healthy usage norms and privacy considerations amidst rapid technological shifts.

Success lies not in technology itself but in how solutions are judiciously designed, aligned, and scaled based on pedagogical priorities – while addressing inequities responsibly. By maintaining a reasonable balance, technology and education can nurture each other potently. Uncontrolled overuse, however, risks losing the wisdom of generations past.

In essence, while transforming 21st-century education fundamentally, appropriate regulation and customization of technology based on developmental needs and learning objectives remain key. Moderation and balance are crucial guiding principles to realize the promise while overcoming the pitfalls on the road ahead.[2]

The Pervasiveness of Technology: Impact on Young Learners

The pervasiveness of technology in education, as discussed above, refers to the widespread integration and influence of technology across various aspects of the educational landscape. It signifies the extensive use of digital tools, devices, and platforms to enhance teaching and learning experiences at all levels of education. This concept acknowledges that technology is no longer confined to a specific subject or classroom but has become an integral and ubiquitous component of the entire education ecosystem.

Artificial intelligence (AI) will only increase the speed, depth, complexity, and benefits of technology's influence and impact in the classroom and on students.

Such benefits include, but certainly are not limited to, ubiquitous access to information and enabling personalized, adaptive, and flexible learning experiences. It facilitates collaborative learning, connects students globally, and supports data-driven instruction. Integrating technology throughout the learning experience equips students with digital skills and prepares them to meet the demands of technology-driven careers.

With benefits also come risks. Technology is essential both in education and for education. AI will only burrow deeper and deeper into daily in-class student exercises and schoolwork, external assignments, as well as assisting educators and administrative functions. Addressing the inherent risks associated with technology is essential to both protecting young learners and preparing them for future roles in society.

The Pervasiveness of Technology: Inherent Risks

The pervasive integration of technology in education has transformed the way we learn and teach, offering many benefits such as better access, interactive learning experiences, and enhanced collaboration. However, this widespread adoption also brings inherent risks that must be considered.

One of the primary concerns is the potential for a widening of the educational divide. As technology becomes more common in classrooms, students of disadvantaged backgrounds may experience a greater lack of access to basic technology and reliable Internet connections, decreasing access to broader educational opportunities. An additional and growing risk is the threat to students' privacy.

The accumulation and retention of vast amounts of digital information about students, from personal information to academic performance indicators, raises concerns about the processing, protection, and long-term retention of these data. Inadequate cybersecurity measures within a school's (or district's) IT system could lead to potential data breaches, leading to a compromise of sensitive student data, which in turn exposes the data owners to various cyber risks (a loss of privacy, for example). Technology can expose children to harmful behaviors. The most common of these is cyberbullying (see Chapter 1).

Excessive screen time and pressure to constantly interact with digital devices can contribute to stress, anxiety, and a decrease in overall well-being. In addition, in some cases, the amount of time a student spends digitally connected to his/her devices raises concerns about digital addiction and its impact on a student's mental health.

In March 2020, the Pew Research Center asked parents a series of questions about their children under the age of 12 and how they engage with digital technologies.

Of respondents with children younger than five, 12% reported the child uses a desktop or laptop computer. The use of gaming devices follows a similar pattern, with 9% of those with a child aged two or younger reporting that their child uses a gaming device. Parents with a child aged five to eight (59%) or age two or younger (49%), reported that their child engages with a smartphone. Parents with a child aged three to four fall in the middle, with 62% saying their child uses or interacts with a smartphone (see Table 2.1).

The same Pew Research Center study reported that among the 60% of parents who say their child younger than 12 ever uses or interacts with a smartphone, six in ten say their child began engaging with a smartphone before the age of five, including roughly one-third (31%) who say their child began this before age two and 29% who say it started between ages three and four.

Some 26% of parents whose child uses a smartphone say that smartphone engagement began between the ages of five and eight. Almost across the board, there is a strong positive correlation between age and the proportion of children using a device.[3]

In a survey of 888 K-12 educators in the U.S. conducted from January 26 to February 7, 2022, by the EdWeek Research Centre, 80% of the respondents reported that increased screen time worsened children's behavior.[4]

Table 2.1 Children's Engagement with Certain Types of Digital Devices Varies Widely by Age

AGE (YEARS)	TV (%)	TABLET (%)	SMARTPHONE (%)	DESKTOP/ LAPTOP (%)	GAMING (%)
0–2	74	35	49	12	9
3–4	90	64	62	21	25
5–8	93	81	59	54	58
9–11	91	78	67	73	68

The rapid pace of technological progress makes it challenging for educators to stay up to date with the evolving tools and platforms. Inadequate training of teachers may obstruct the effective integration of technology into the curriculum and limit its potential benefits. Furthermore, reliance on technology in education can lead to devaluing traditional teaching methods and interpersonal skills.

The proliferation of "Personalized Learning" may lead to an unbalanced delivery of education where silicon-based instructors instead of carbon-based instructors are the primary in-class educators. This may also be cost- and budget-driven as school administrators seek to replace the more expensive (salary, benefits, etc.) teachers with less expensive (amortized over many years) technology.

Students may become overly dependent on digital resources, potentially hindering their ability to think critically, solve problems independently, and communicate effectively in face-to-face situations. Additionally, as discussed in Chapter 1, technology (in the form of AI, for example) is not always correct.

The need to learn and think independently is a critical skill for learners of all ages. To recap, from Chapter 1, the risks associated with AI that students should be made aware of include:

- Lack of Transparency: Because of the complexity of the design and functionality of AI systems, it can be difficult for humans to understand how an AI system has come to a particular decision or recommendation.
- Data Privacy: The huge amounts of data used to train AI systems can lead to the exposure of private, personal, and sensitive information.
- Vulnerabilities in AI Systems: AI systems can be exposed to attacks just like any other computer system.
- Misuse of AI: AI can be used for malicious purposes to say mean and untrue things about people. AI can even create completely inaccurate stories.

Furthermore, there is a risk of overemphasizing standardized testing assessments and data-driven metrics in the technology-driven education landscape. The focus on quantifiable results may neglect the development of essential skills such as creativity, critical thinking, and emotional intelligence, which are crucial for success in a rapidly changing world. Lastly, the susceptibility of technology to glitches, outages, and technical issues can disrupt the learning process and create challenges for both educators and students. Reliance on digital platforms without robust backup plans can lead to significant disruptions in the educational environment.

While the integration of technology in education has the potential to revolutionize learning experiences, it is crucial to acknowledge and address the inherent risks associated with its pervasive use. Balancing the benefits of technology with concerns related to accessibility, privacy, mental health, teacher training, skill development, and technical reliability is essential to creating an educational environment that fosters holistic growth and prepares students for the challenges of the future.

Importance of Cyber Safety Education — Why Teach Kids about Cybersecurity?

Hanover Research's 2023 Trends in K-12 Education report highlights both new and ongoing issues and priorities that are anticipated to affect K-12 programs in 2023 and presumedly beyond. The report identifies five specific trends that will shape K-12 education. Of specific interest (although all trends identified are of interest) is trend number 5, "Protecting Student Well-Being Demands Systemic Support."

Hanover states that this trend is "to address the 'whole child,' including their social, emotional, mental, and physical needs, schools embrace a systemic approach that targets all elements of student wellness."[5]

Digital technology is rapidly changing the way we educate and protect students in our classrooms. While the Internet can be a tremendous resource for young learners, it also brings a host of potential risks, from cyberbullying to viewing hurtful and sensitive content.

The U.S. Department of Education and the U.S. Department of Health and Human Services identified, in the department's joint report, Policy Brief on Early Learning and Use of Technology, four guiding principles for the use of technology with early learners.

These four guiding principles are as follows:

- Guiding Principle #1: Technology—when used appropriately—can be a tool for learning.
- Guiding Principle #2: Technology should be used to increase access to learning opportunities for all children.
- Guiding Principle #3: Technology may be used to strengthen relationships among parents, families, early educators, and young children.
- Guiding Principle #4: Technology is more effective for learning when adults and peers interact or co-view with young children.[6]

It is interesting to note some surprising, if not concerning, information revealed in the Annual Cybersecurity Attitudes and Behaviors Report 2023, sponsored by the National Cybersecurity Alliance and CybSafe, when asked to respond to the question, "I feel that staying secure online is worth the effort …"

> Of participants responding, 69 percent thought staying secure online was worth the effort. But the younger generations (21% of Gen Z and 23% of Millennials) are skeptical about the return on investment. They were more than twice as likely as Baby Boomers (6%) and the Silent Generation (9%) to doubt online security is worth the effort.[7]

Although technology or the pervasiveness of technology is not specifically addressed in the Hanover Trends report, or the Department of Education and the Department of Health and Human Services report, with the continual evolution of technology and the accelerating use of technology in classrooms and by students, it is imperative that educating children on the responsible use of technology and keeping children safe in the cyber world in which they learn, play, and reside should be a priority supported by academic administrators and an important objective of K-12 educators.

In essence, technology has transformed teaching and learning by becoming a fundamental, integrated component of education.

Cybersecurity Topics for Young Learners

Why are the findings from Hanover Research's 2023 Trends in K-12 Education report of concern when we examine methods to keep students cyber-safe? If not taught to recognize the risks and to take prudent steps to be cyber-safe while using technology, students could potentially place themselves, as well as their family, and their friends at risk, or in a worse case, in danger.

As an example, many web pages are designed to look like legitimate sites that students should trust. As a teaching exercise, educators may wish to address the risks that students may face from this tactic, which include but are not limited to:

- Inappropriate Content Exposure: Fake websites may host inappropriate content that is not suitable for children, including violence, explicit language, or graphic imagery. Children might accidentally stumble upon these sites, leading to potential harm or exposure to content that is not age-appropriate.
- Online Predators: Malicious actors may create fake websites to lure children into sharing personal information or engaging in unsafe online behavior. These predators can pose as friends or trustworthy figures, putting children at risk of exploitation or harm. Read more on this topic in Chapter 3.

  When talking about "malicious actors" with young learners, try this approach:

  Malicious actors are like tricky characters on the Internet who try to do unkind things. They might pretend to be someone they're not, or they could try to make computers sick, so we need to be careful and smart when we're online.
- Phishing Scams: Fake websites often use deceptive tactics to trick kids into sharing sensitive information, such as passwords or personal details. Children may unknowingly fall victim to phishing scams, compromising their online security.

  When explaining "phishing scams" to young learners, try using the following:

  Phishing scams are like pretend messages or emails that want to trick us into sharing our secrets. Just like we don't share our special passwords with people we don't know, we need to be extra careful and not click on links or share information with emails that seem a bit fishy.

- Malware and Viruses: Phony websites can be a breeding ground for malware and viruses. Clicking on links or downloading content from these sites may expose children's devices to harmful software that can damage or compromise their digital devices.

  You may wish to try this approach when discussing malware and viruses with young learners:

  Just like you can get sick and catch a cold, malware is like a naughty bug for computers that can make them sick. It's important to avoid clicking on strange things and only visit approved websites to keep our computers healthy and happy.

  Computer viruses are like little germs that can make our computers feel unwell. They sneak in when we click on things we shouldn't, so we need to be good computer doctors and only open things from people we know and trust.
- Identity Theft: Children may not fully grasp the concept of identity theft, making them vulnerable to fake websites that collect personal information. This can lead to the unauthorized use of their identity for malicious purposes.
- Cyberbullying: Bogus websites may be used as platforms for cyberbullying, where children can be targeted, harassed, or humiliated by others. This can have serious emotional and psychological consequences for the child involved. This topic was addressed in greater detail in Chapter 1.
- Financial Scams: Some fake websites may attempt to trick children into making unauthorized purchases or divulging their parents' financial information. This poses a risk of financial loss and potential legal consequences.

  When reviewing the topic of technology-based financial scams with young learners, you may try approaching the subject in the following manner…

  Financial scams are like tricky games that some people play to try and take away our money without permission. It's important to be money detectives, always checking with grown-ups before sharing any money information online or in person.

- False Information: Children may encounter fake websites that present false information, which can mislead them in their learning or understanding of various subjects. This misinformation can have educational and cognitive consequences.
- Unsupervised Online Interactions: Phony websites might encourage children to engage in unsupervised online interactions, such as chat rooms or forums, where they could be exposed to inappropriate discussions or interactions with strangers.
- Addiction and Distraction: Fake websites designed to resemble popular games or entertainment platforms may contribute to excessive screen time, leading to potential issues of addiction and distraction from essential activities like homework, physical activities, and family time.

  When explaining "fake websites" to young learners, try this approach:

  Fake websites are like pretend playgrounds on the Internet that might look fun but can be a little bit tricky. Just as we choose safe and real playgrounds to play in, we should ask grown-ups for help to finding safe and trustworthy websites when we explore the Internet.

Given the growing level of risk in today's online environment, educators should use the opportunity whenever possible via classroom exercises and school assignments to communicate the importance of staying cyber-safe to young learners. Parents and trusted adults can assist educators by providing practical examples and role-playing scenarios to help children understand the importance of cybersecurity. See the lesson plans provided in this chapter for some samples and suggestions.

Connected and Protected: Instilling Smart Habits in Young Tech Users

The unavoidable impact of technology on our personal lives is a defining characteristic of our current digital society and the technological environment in which we live and work. From Smart Home devices that regulate temperature and lighting to virtual assistants that respond to voice commands, technology is seamlessly integrated into almost every aspect of our daily lives.

While technology and continual technological advancements offer convenience and efficiency, they also raise important considerations regarding privacy. Our constant connectivity to devices (mobile phones, tablets, laptops, TVs, etc.) and the data collected for analysis via these devices pose considerable risks to our privacy. Such risks warrant a continual assessment of the importance of balancing the benefits of technological innovation against the loss of personal privacy and an overdependency on the technology itself.

As our homes become increasingly interconnected, the need to evaluate the ethical use of technology within private spaces (i.e., homes) becomes vital to ensuring a balanced coexistence between digital and personal domains.

Students will need to function, manage, and learn how to remain cyber-safe as omnipresent and all-consuming technology continues to seep into their personal lives and even their own homes. Teaching children about the safety aspects of Smart Home technology (including the risks and benefits associated with these technologies) is critically important.

Smart Home technology involves the integration of devices and systems within a home that can connect to the Internet, enabling users to control and automate various functions remotely. These devices, such as smart thermostats, lights, appliances, entertainment systems, motion sensors, indoor (and exterior) security cameras, and voice-activated assistants, can be managed through a central hub or mobile app. Through this connectivity, users can customize and monitor their home environment, enhancing convenience, energy efficiency, and overall security.

Smart Home technology aims to create an interconnected and automated living space that responds intelligently to the preferences and needs of its inhabitants. The aim is to enhance efficiency, convenience, and security by enabling users to remotely manage and automate functions within their homes.

The Internet of Things (IoT) refers to a network of interconnected devices, objects, or "things" embedded with sensors, software, and other technologies, enabling them to collect and exchange data.

These devices, ranging from everyday objects like thermostats and appliances to more complex systems like wearable devices, can communicate with each other through the Internet, facilitating data sharing and automation. The goal of IoT is to enhance efficiency, provide new functionalities, and improve overall connectivity in various aspects of daily life.

Interconnected Living: Exploring the Internet of Things in Smart Homes

IoT and Smart Home technology share similarities but differ in scope and application. Both involve the integration of devices with Internet connectivity, allowing for communication and data exchange. However, their primary distinctions lie in their broader purpose and specific focus.

IoT and Smart Home technology involve connecting devices to the Internet for improved functionality, IoT has a broader, industry-spanning scope, whereas Smart Home technology is a subset of IoT, focusing specifically on enhancing residential living spaces. It includes devices like smart thermostats, lights, security cameras, and voice-activated assistants designed to improve convenience and efficiency within homes.

IoT and Smart Home technology introduce specific risks to young learners and children, primarily centered around privacy, security, and potential exposure to inappropriate content. These interconnected devices often collect and share data without the user's knowledge, raising concerns about the privacy and possible use of children's private information. The risk of unauthorized access to personal details and habits poses a threat to the confidentiality of children's data, potentially leading to issues like identity theft or misuse. Additionally, the integration of cameras and microphones into Smart Home devices raises concerns about the unintentional capture of private moments, again impacting children's right to privacy.

The risks and security vulnerabilities in IoT and Smart Home technology may expose children to cyber threats, including hacking and phishing attempts. Unauthorized access to connected devices can lead to various risks, such as surveillance, unauthorized control of Smart Home functions, or the manipulation of devices to expose children to inappropriate content. Furthermore, the increasing dependency on these technologies may contribute to issues like excessive screen time and hinder the development of critical thinking skills as children engage in automated and interconnected environments (read more on this risk and its relationship to cyber-safe practices later in this chapter).

To address these risks, educators must stay informed, discuss with students the need to implement strong privacy settings when engaging with technology, and guide children in an overall approach to responsible Internet use. Discussions and class exercises with students can be developed to address the importance of establishing regular communication with a trusted adult. With trusted adults, students can learn to establish appropriate boundaries, which contribute to creating a safer digital environment.

The next section examines digital citizenship and continues to reinforce concepts discussed in Chapter 1 on the importance of character development and that good character determines good behavior. A good digital citizen respects others, speaks up when they see something inappropriate, hurtful, or dangerous, and protects themselves and their personal information.

Digital Citizenship

Defining Digital Citizenship

Digital citizenship is the ethical, moral, and responsible use of technology to ensure one's own and others' protection while collaborating in an increasingly digital, networked, and global society.

Because of the unique issues that technology has brought into our daily lives, everyone must learn about digital citizenship so that each member of society can become aware of the dangers and pitfalls as well as the positive outcomes associated with taking on the role of a digital citizen in a global community. Although citizenship may be rooted in similar foundational contexts in both offline and online environments, digital citizenship yields many special case issues that must be considered to elicit appropriate and responsible actions in online settings.

For students, the tenants of digital citizenship are far-reaching, as the social media technologies with which they are intimately familiar imply several consequences that can have serious implications for their personal, educational, and future business lives.[8]

Expanding this definition to be more inclusive, the concept of digital citizenship grows to encompass a range of skills and literacies that can include Internet safety, privacy and security, cyberbullying, online reputation management, communication skills, information literacy, and creative credit and copyright.[9]

What or Who Is a Digital Citizen?

A digital citizen is an individual who engages responsibly, ethically, and positively in the digital world.

Being a digital citizen involves using technology, especially the Internet, in a manner that respects the rights and well-being of others, while also contributing to the overall betterment of the online community.

Digital citizenship encompasses a range of skills, knowledge, and attitudes that empower individuals to navigate the digital landscape safely and effectively. This includes understanding and practicing concepts such as online etiquette, responsible use of information, digital literacy, and awareness of potential risks and challenges in the digital environment.[10] Becoming a good digital citizen begins with developing, fostering, and sustaining good character and acceptable behaviors, as was discussed in Chapter 1.

Our Digital Playground: Learning to Be Safe and Kind Online

Teaching children to be good digital citizens and to **ABIDE** by simple yet effective principles will help them to develop a cyber-safe approach to actively living, learning, and safely participating in an ever-evolving digitally infused, digitally dependent society.

The following **ABIDE** principles may be presented to your students when discussing how to become a good digital citizen.

Be **Attentive**

Children strive to accomplish this goal by demonstrating a heightened awareness and careful consideration of one's digital actions, interactions, and online surroundings. When kids are "attentive" online, it means they're paying close attention to how they act online, being careful not to do anything risky or hurtful, and making the Internet a better and safer place for everyone. Being attentive online involves a mindful approach and understanding how the consequences of one's actions can affect others.

Be **Broadminded**

Young learners who are "broadminded" are more apt to consider hearing different points of view, are more willing to respect the opinions of others, and engage in online interactions with empathy and understanding. Being broadminded means students respect and appreciate how everyone is different, which helps them to remain open-minded when working and playing in an online environment. A young learner who exhibits a broadminded approach to life will value diversity and contribute positively to personal and online interactions with peers.

Become **Immersed**

When children become fully engaged and involved in using technology to solve problems there is a positive impact on learning new skills and acquiring knowledge, they find creative solutions to solve problems and provide a positive contribution to their social environment.

Be **Discerning**

To a young learner, this is about how good the student is at checking if what they see online is true and reliable. Thinking carefully, asking questions, and figuring out exactly what information is trustworthy, and which might be wrong are learned skills. Through teaching and demonstration, children will learn how information can be biased, notice when that happens, and know when and how to double-check information to make sure it's from a reliable source. An important step in becoming and remaining cyber-safe in today's online environment.

Be **Evenhanded**

Teaching students to be "evenhanded" is about helping them find the right balance between their online and "offline" real-life environments, the balance between online activities (e.g., games and movies) and offline activities (e.g., spending time with friends and playing without screens). Students who act in an evenhanded manner when it comes to the digital world will spend their time wisely, be balanced between online and real-life environments, and have a good time in both environments. In the end, it is all about finding the best mix of both!

Growing Up Digital: A Summary

Discussing the concept of digital citizenship and what it means to be a good digital citizen with students helps to prepare them for the world that awaits them as adults.

A part of guiding young learners toward becoming good digital citizens is to empower them to handle issues that they may encounter when using technology.

To summarize and provide an approach to discussing this topic with your students, we recommend the following:

- Explain to students the public nature of the Internet and its risks and benefits. Be sure they know that any digital information they share, such as emails, photos, or videos, can easily be copied and pasted elsewhere and is almost impossible to take back. Remind students that some of this digital communication, like social media posts or photos, could damage their reputation and friendships and should not be shared.
- Remind students to be good "digital friends" by respecting the personal information of friends and family and not sharing anything online about others that could be embarrassing or hurtful.
- Students may face situations like cyberbullying, unwanted contact, or hurtful comments online. Work with them on strategies for when problems arise. These can include talking to a trusted adult right away, refusing to retaliate, calmly talking with the bully, blocking the person, or filing a complaint. Agree on steps to take if the strategy fails. It is better to have these strategies in place ahead of time and be proactive instead of being reactive after an event has occurred.

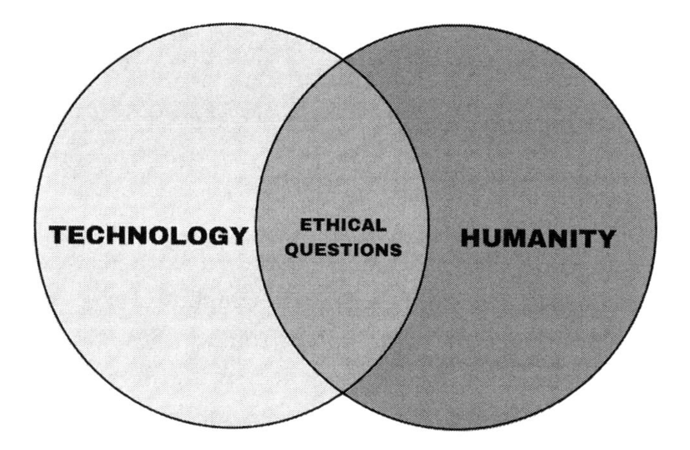

Figure 2.1 Good digital citizens have acquired the necessary skills to address ethical questions at the intersection of technology and humanity.[13]

Teaching digital citizenship to students at a young age helps children hone effective communication skills, allows them to practice good form in social participation, and protects themselves online. It's crucial to learn early about showing respect for others online, recognizing threats, and developing habits that take advantage of the benefits of the Internet while creating resilience to its dangers.[11]

The final word, as the International Society for Technology in Education (ISTE) summarizes so nicely, is "Digital citizenship is about more than online safety. It's about creating thoughtful, empathetic digital citizens who can wrestle with important ethical questions at the intersection of technology and humanity" (Figure 2.1).[12]

Basic Cybersecurity Concepts for Young Students

Understanding Personal Information

To begin with, broadly, personal information is any information that can be used to distinguish or trace an individual's identity, either alone or when combined with other information that is linked to or linkable to a specific individual.

Conveying this concept to young learners should be done in a simple and relatable manner. For example…

> Hey everyone! Today, we're going to talk about something really important – our personal information. Just like you have special toys

> or snacks that are just for you, personal information is your special stuff that you keep safe.
>
> Personal information is things about you that are special and private. It's like your own treasure! Let's talk about a few personal things.

What Is Personal Information?

Naturally, personal information will vary in both its breadth and depth based on an individual's age. There are, however, several items/identifiers that universally represent personal information. Teaching young learners to recognize these identifiers and how to keep them safe is a critically important step in developing a cyber-safe approach to engaging in and living in a technology-dependent society. A dependency that only grows stronger daily!

What are the most elementary and important personal information identifiers that students should be taught to protect, and how can this information be communicated to young learners?

Consider the following identifiers and an approach to opening discussions with your students...

> Personal information is special because it helps keep us safe. Just like we don't give our favorite toys to strangers, we don't share our personal information with people we don't know well. It's our way of making sure we are safe and happy.

Name

> Your name is super special, like a magic word that belongs only to you. We don't share our names with just anyone; we share them with people we know and trust.

Address

> Your home address is like the secret code to where you live. It's something only your family and some special friends should know. We keep it safe, just like a hidden treasure. Consider using the analogy of a treasure map, where only certain people can know the way to their special treasure (home).

Phone Number

> Your phone number is like a special song that only you and your family sing. It helps you talk to the people you love. We don't sing this song to everyone, just our family and very close friends. Help students make

up a simple tune or rhythm to help them remember the sequence of numbers.

Family and Friends

Personal information also includes things about your family and friends. We keep our family's information safe, just like we keep our information safe. Use examples of people they know and trust, like parents, grandparents, or teachers.

Online Games and Apps

When we play games or use apps, sometimes they ask for information. But we only share that information with Mom, Dad, or a trusted adult. It's like asking for permission before playing a new game.

So, remember, if someone, even if they seem nice, asks for your personal information, always check with a grown-up you trust. They will help you decide if it's okay or not. Remember, personal information is like a secret treasure, and we want to keep it safe!

This list is representative, additional identifiers may be added as appropriate based on student interaction, response, and feedback received by the teacher throughout the class discussion.

Digital Imprint or Digital Footprint?

Before discussing digital footprints, let us take a moment to briefly look at what is called your digital imprint. The terms "digital imprint" and "digital footprint" are often used interchangeably, but they can have slightly different connotations depending on the context.

The term "digital imprint" is less commonly used and can be interpreted in a broader sense. It can refer to the entirety of one's online presence, including both intentional and unintentional elements. A digital imprint could encompass not only the active digital footprints generated by intentional online actions but also the passive elements, such as the information collected by websites and online services without explicit user input.

The concept of a digital imprint might be used more broadly to describe the comprehensive online identity or the collective impact of a person's digital activities.

What's a Digital Footprint?

A digital footprint refers to the trail, traces, or records of a person's online activities. It encompasses the data and information that individuals leave behind while using digital devices, or services, or whenever someone posts information about you on social media platforms (e.g., Facebook, YouTube, TikTok, Snapchat, X, Pinterest, LinkedIn, etc.). Digital footprints include the websites visited, social media interactions, online searches, posts, comments, and any other online actions that create a traceable record. Digital footprints are often discussed in the context of online privacy and reputation management. They highlight the lasting impact and visibility of one's online actions.

Having a digital footprint in today's digitally-infused society is hard to avoid and normal. The concern is that your digital footprint and all the related personal information associated with that footprint are publicly available.

What types of digital footprints may young learners leave on the digital landscape? What is the best approach to discussing with students how to act in a cyber-safe manner to limit their footprints when using technology? These questions and other cyber-safe habits for young learners are discussed in the next section.

Digital Footprints and Young Learners

Digital footprints and the risks of leaving such footprints in the digital world for everyone to see are important concepts for young learners to understand. This understanding will help them increase their cyber safety awareness skills and help mitigate the associated risks when interacting with technology. Skills, which will be important as they learn to navigate their way through the ever-evolving and evasive world of technology in which they live.

Digital footprints are created in two ways: passively and actively.

A passive footprint is created when your data is collected, usually without your being aware of it. Common examples are search engines storing your search history whenever you're logged in, and web servers logging your computer's IP address when you visit a website.[14]

To discuss the concept of passive footprints with students, consider using the following examples of where students may leave passive digital footprints and how to be cyber-safe in the digital world.

- Online Games and Apps
  Passive Footprint: When playing games or using apps, data about how long a student plays, their favorite activities, or even their location may be collected without them knowing.
  Cyber-safe: Stress the importance of privacy (as discussed previously) and the need to make sure that privacy settings are in place before accessing and playing games. Parents can help too by choosing apps with strong privacy settings and explaining the importance of asking for permission before playing new games. Encourage kids to play offline games as well.
- Smart Toys
  Passive Footprint: Interactive toys may record conversations or collect data on a child's preferences to enhance their play experience. It's important to be aware of potential privacy and security concerns. These concerns often stem from the toy's ability to connect to the Internet, collect personal data, or interact with children in ways that could compromise privacy.
  Smart Toy Examples:
  Hello Barbie: This interactive doll uses speech recognition technology to have real-time conversations with children. Concerns arise from the fact that the doll records and transmits conversations to a server for processing, raising questions about data security and privacy.
  Cayla Doll: Similar to Hello Barbie, Cayla is an interactive doll that can converse with children. It connects to the Internet via Bluetooth. The cyber safety issue is a lack of secure connections, potentially allowing unauthorized access to the doll and the data it collects.
  Furby Connect: This updated version of the classic Furby toy uses Bluetooth to connect to an app, through which it updates and receives new content. The Bluetooth connection has been noted as potentially insecure, making the toy vulnerable to hacking.

Educators should discuss these potential exposures with students. Additionally, parents and guardians need to research and understand the privacy and security features of smart toys before introducing them to children. This includes understanding data collection practices, the security of Internet or Bluetooth connections, and the company's track record for handling personal information.

Cyber-safe: A discussion with young learners could proceed as follows....

Remember, when we play with smart toys that can talk or connect to the Internet, we want to make sure they are safe, just like we do when we cross the street with a grown-up. So, we always check to make sure they have their 'safety belt' on, which means they have a secret code that only you and your family know. And just like we keep our name, home address, and phone number private at the playground, we don't tell thesc things to our smart toys during playtime.

- School Computers and Websites

Passive Footprint: Discuss with your students how their school computers and educational websites, which they may visit, could track their progress and activities. Some of this collected data may be used for academic or product assessment purposes.

Cyber-safe: Discuss with your students the legitimate purpose of software tools that collect information about the sites they visit when they are online and reassure them that their data is used to enhance learning. Emphasize the importance of keeping login information private. Explain that sometimes leaving footprints is alright, unavoidable, and necessary, if done so with the student's knowledge and approval. Once again, revisit the concepts of privacy and personal information.

You may wish to approach this topic through a discussion with your students that could proceed as follows....

In the online world, like when you use a computer or a tablet to play games or watch videos, you also leave behind "digital footprints." These are like tiny clues about what you like to do online. For example, if you watch a lot of videos about dinosaurs, the computer learns that you like dinosaurs and might show you more dinosaur videos.

Data collection is like someone collecting all these digital footprints. They gather this information to understand what you like and don't like. This can help them make the games and videos you see even more fun for you. But just like in the playground, it's important that someone responsible is taking care of you. That's why your parents or teachers set rules for how you use the Internet to keep your digital footprints safe and private.

- Photo Sharing

  Passive Footprint: When parents share pictures of their kids on social media, information about the child's interests and activities may be collected. Talk with students about reminding their parents or a trusted adult about the risks of sharing private information.

  Cyber-safe: Educators should be mindful of what they may share online about their classroom activities and adjust their privacy settings accordingly. Teach kids about the importance of not sharing personal details online. Your actions can be the best cues for acting cyber-safe when using technology. Your young learners will want to copy your actions!

- Virtual Assistants

  Passive Footprint: Discuss with students being alert and aware of devices like smart speakers that may record conversations to improve their (the smart speaker's) understanding of user preferences.

  Smart Speaker Examples:

  These devices often come with voice recognition features that allow for interactive play but may raise privacy concerns due to their connectivity, data handling practices, and capability of recording and potentially transmitting conversations to cloud-based servers.

Amazon Echo Dot Kids Edition: This device is designed with kids in mind and can play music, answer questions, read stories, and more. Conversations can be recorded and processed on Amazon's servers to improve AI responses.

CogniToys Dino: This educational toy uses IBM's Watson technology to engage with kids in conversations. It learns from the child's interactions and can answer a wide range of questions, which involves processing the conversations through cloud-based servers.

Mattel's Hello Barbie: This smart doll uses speech recognition and connects to Wi-Fi to engage in two-way conversations. The audio recordings of these interactions are sent to cloud servers for processing to generate responses.

Fisher-Price Smart Toy Bear: This interactive plush toy uses voice recognition and an accompanying app to engage in storytelling and games. Conversations with the toy can be uploaded to the cloud to tailor the learning experience.

Cyber-safe: Smart speakers and the role they play in a child's toy may be new and surprising to some students. You may wish to approach a discussion of this topic with young learners as follows…

Imagine you have a toy that can listen and talk to you. When you want to talk to it, you say, "Hello, toy!" and then ask it to do something like play music or tell you a story. But sometimes, when you're done playing, the toy might still listen to you, even if you're not talking to it. It is like they have ears that are always listening.

To make sure the toy doesn't listen when you don't want it to, you can turn off its ears. It's like putting it to sleep. You or a grown-up can usually find a little button on the toy or in its settings—that's like the toy's secret off switch for its ears. When you turn it off, you can be sure it's not listening anymore.

And remember, when the toy's ears are on, you should be careful about what you say. It's like if you were telling secrets to your friend, you wouldn't want someone else to hear them. So, if you're talking about secret stuff or private things, it's a good idea to make sure your toy's ears are off. That way, you can keep your secrets safe and only share them with those you want to. Always ask a teacher or trusted adult if you're not sure how to turn the ears off or if you want to make sure your toy is sleeping and not listening.

An active digital footprint is created when you voluntarily share information online. Every time you send an email, publish a blog, sign up for a newsletter, or post something on social media, you're actively contributing to your digital footprint.[15]

To explore the idea of active digital footprints with students. Introduce the concept by providing examples of situations where students may leave active digital traces of their online activities and providing guidance on practicing cyber safety in the digital world.

For example....

- Sharing photos or videos
  Active Digital Footprint: Uploading pictures or videos to social media or sharing them with friends online.
  Cyber-safe: Encourage students to check with a trusted adult before sharing any photos or videos. Emphasize the importance of obtaining permission and being aware of what they shared and with whom.
- Posting Personal Information
  Active Digital Footprint: Sharing personal details like their full name, home address, the names of brothers or sisters, or their school on websites or apps.
  Cyber-safe: Teach students to keep personal information private and only share it with their parents or trusted adults. Discuss the concept of online privacy and the importance of not revealing too much about themselves.
- Commenting or Chatting Online
  Active Digital Footprint: Engaging in conversations or leaving comments on websites, games, or social media.
  Cyber-safe: Emphasize the significance of being respectful and kind online. Remind students not to share private information in comments or chats and to report anything that makes them uncomfortable.

- Creating Usernames and Passwords
  Active Digital Footprint: Choosing usernames and passwords for online accounts.
  Cyber-safe: Teach the importance of creating strong and unique passwords. Encourage students to use a mix of letters, numbers, and symbols and to keep their passwords private, only sharing them with their parents or a trusted adult. See the following section, "Guardians of Access the Crucial Role of Passwords," for more information to share with your students on creating cyber-safe passwords.
- Online Learning Platforms
  Active Digital Footprint: Participating in online educational activities and submitting assignments.
  Cyber-safe: Students should be taught to follow the teacher's guidelines for online interactions. Reinforce the concept that the Internet is a place for both learning and fun. Remind them that the Internet can also be risky and that it is important to remember and practice their cyber safety habits when exploring the digital world.

Creating analogies between the physical footprints animals leave and the digital footprints that students create and leave behind in a digital environment can help make explaining the concept of digital footprints more relatable to your students.

Here are several examples.

- Digital Footprint: Social Media Posts
  Analogy: Animal tracks: Lion paw prints in the sand (see Figure 2.2).
  Explanation: Just as a lion leaves paw prints when walking that can tell which direction the lion has come from and in what direction he is going, anything you post on social media sites creates a digital trail. Your posts are like the lion's paw prints and can be connected back to you and those with whom you share them, forming a digital footprint.

Figure 2.2 A lion's paw prints in the sand.[16]

- Digital Footprint: Online Comments
  Analogy: Cat footprints in the snow (see Figure 2.3).
  Explanation: A cat, a dog, or even a bear, leaves tracks when walking in the snow. The animal's tracks can be used to identify the type of animal that left them. In the same way, you can be identified through your online comments, which can be traced back to you. Posting online comments creates a digital trail of your activities.
- Digital Footprint: Sharing Personal Information
  Analogy: Animals leave scent marks.
  Explanation: Just as some animals leave scent marks to communicate information about themselves to other animals, sharing personal details online creates a "digital scent" that other people can follow and know that it is you. Therefore,

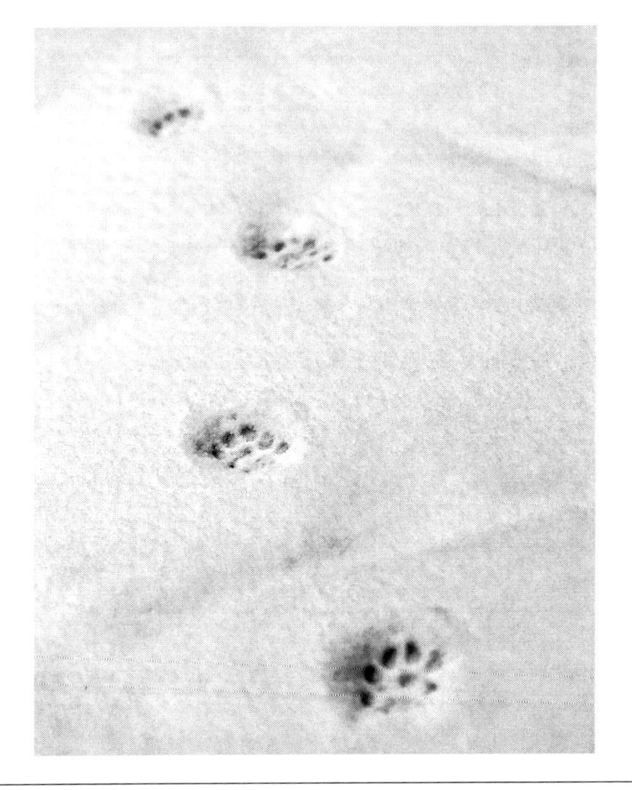

Figure 2.3 Cat footprints in the snow.[17]

it is important to be very careful about what information you share and with whom.

The concept of "scent marks" may be a bit more of a challenge to explain to young learners. The following is an approach that may succeed in conveying the concept.

Ask the students to imagine that they have a special way of leaving a message for their friends to find, to let them know that they were waiting in the classroom but left to go outside to the playground.

Explain that all animals have their own special way of leaving messages for other animals to find. Just like you would write a note or draw a picture for your friends, animals sometimes leave special smells called scent marks. It is the animal's way of saying, "I was here, and here is a little smell to say hello!"

Figure 2.4 Puppy leaving scent marks.[18]

For example, imagine a puppy walking through a yard. Just by walking through the yard, the puppy leaves tiny scent marks where he walks. These scent marks are like a secret message that only another animal with a very good sense of smell can understand. Each animal has its own special smell (scent), and they use it to leave messages for other animals (see Figure 2.4).

- Digital Footprint: Online Gaming/Searching/Social Media
  Analogy: Animal traces of feathers or fur (see Figure 2.5)
  Explanation: When animals walk, run, or fly through an area, they might leave traces of feathers or fur behind. Just the same way as you scroll online or engage in online activities, you leave traces or records of your actions (or behaviors) that can be linked right back to you, forming a digital trail of your online habits and activities.

Figure 2.5 Bird feather.[19]

- Digital Footprint: Email Communication
  Analogy: Animal camouflage (see Figure 2.6).
  Animals protect themselves by using the color of their fur or skin, through patterns, shapes, stripes, spots, and other means of natural camouflage, to blend in and hide from predators. Emails that you may send will create a trace or trail specific to your identity and personal activities. Each email you send contributes to creating an overall

Figure 2.6 Animal camouflage.[20]

> personal digital footprint that is uniquely yours! Learn to camouflage yourself by taking steps to know who you are emailing; don't send any personal or private information to anyone you don't know; and ask a parent or trusted adult before opening or responding to an email from someone you do not know.

By using recognizable analogies between digital footprints and animal footprints/characteristics/behaviors, students will recognize and understand that their online activities, like the actions of animals in the natural world, leave behind traces that can be tracked and associated with them. This analogy exercise helps simplify the concept of digital footprints, making it easier for young learners to grasp and understand.

A Quick Note Regarding Browser Fingerprinting for the Educator

Browser fingerprinting is the act of gathering data from a web browser to create a device's (laptop, mobile phone, etc.) unique fingerprint. This procedure can make remarkable amounts of information about a user's software and hardware environment available, and it can be used to create a special identity termed a browser fingerprint. An analogy with a digital fingerprint is appropriate as the browser fingerprint is often unique to an individual user.

Browser fingerprinting has, without a doubt, become a significant privacy threat to web users. When a user visits a website, browser fingerprinting is used to gather information about system settings and browser configurations, the type and version of the browser, as well as the operating system, IP address, and other current system/user settings.

In the end, a unique identity can be created using this procedure, which can also expose a surprising amount of information about a user's software and hardware settings. The privacy implications are significant because users can then be tracked using these fingerprints.

A third party can identify a person and link their browsing behaviors within and between sessions by gathering browser fingerprints. Most significantly, because the tracking scripts are silent and run in the background, the user is oblivious to the data collection process, which is fully transparent.

Even when the user disables the use of cookies by websites, fingerprints can be used to identify particular users or devices, fully or even partially. Many users mistakenly think cookies are browser fingerprints; they are different. The difference is that cookies are only accessible to the website from which they were obtained and cannot be shared from one website to another, whereas browser fingerprinting can track users throughout the Internet.[21]

Discussing Browser Fingerprinting with Young Learners

The topic of browser fingerprinting is important enough that it should be discussed with young learners. The discussion, however, will require a simplified, non-technical approach. The following is a suggested approach that educators may use when discussing browser fingerprinting.

Ask students to imagine that they have a "magic paintbrush" (see Figure 2.7). Their "magic paintbrush" is special, one-of-a-kind, the only one like theirs in the world, and it remembers what colors they like to paint with. Every time they use their "magic paintbrush," they can always paint with their favorite colors.

Explain to students that just like their "magic paintbrush" that remembers what colors they like, a computer, tablet, or even a mobile phone can remember their favorite websites or games. Remind students to be cyber-safe when online, keep their private information private, and if they ever feel uncomfortable with anything they see online, tell the teacher or a trusted person.

Figure 2.7 Magic paintbrushes.[22]

Guardians of Access: The Crucial Role of Passwords

Passwords are a first line of defense in protecting the information that is important to you and restricting individuals from accessing data (or information) that they are not authorized to have access to.

Having explained to students the importance of keeping cyber-safe and their personal information private, discussing passwords and the role passwords play in a digital world is an important lesson.

While the subject of passwords and their role in securing our computing devices and the data that reside on them can be technical, explaining the basic concepts to young learners will require a simpler approach.

Have students imagine a room filled with all sorts of fun toys (see Figure 2.8). Tell them that if they know the secret code, they can unlock the door, go inside, and play with the toys. Explain that their password, just like the secret code that opens (unlocks) doors, is used to open (unlock) their tablet, computer, or other electronic devices.

Continue by explaining that it is very important to keep this secret code secret. If someone finds out their secret code and changes it, they will not be able to unlock (open) their tablet or other electronic

Figure 2.8 Room filled with toys.[23]

devices. Stress that secret codes are to be kept secret, and do not tell or give your secret code to anyone unless your teacher, parent, or a trusted person asks you.

Importance of Strong and Unique Passwords

Now that your students know what a password is, how it works, and that it should be kept secret, discussing how to create a secure password is our next step.

The longer and more complex a password is, the more difficult it will be for someone to randomly guess it. If the password is complex, the interloper will move on to an easier target. Thus, the objective is to create a strong, complex password that is neither time- nor cost-effective for someone to attempt to guess or try to bypass.

Great! That works well for most adults (at least those who follow the rules of creating strong passwords or, better yet, passphrases), but what about our young learners?

It is best to begin by explaining that a tricky secret code is better than a simple or easy one and is much more difficult for someone to guess. However, for some young learners, remembering their tricky secret code may present a challenge.

The challenge then becomes how to explain to a student the need to create a password that is both secure and, at the same time, easy to remember. Emphasizing that their secret code and password should be a little tricky so that only they know it, but not so tricky that they forget it.

A suggested approach is to encourage young learners to create a password using pictures or drawings. For example, if their password is “Bluestar7,” they can draw a blue star and the number 7 next to the word. This visual cue helps them remember.

Perhaps some of your students like to play sports, so create a password such as “SoccerChamp22.” Drawing a soccer ball and writing the number 22 next to their password provides an important visual clue. This visual clue encourages them to remember their password.

For the students who may have a favorite book, helping them to create a password that combines the book title and a primary color is an approach to creating a strong but easy-to-remember password.

"RedMatilda," for example, could be a good start. Once again, keeping the language simple, using relatable examples, and involving them in the process of creating their password often leads to successfully memorizing their password, and besides, it is fun!

Reinforce the importance of keeping the password a secret. Explain that just like they don't share their toys with strangers, they shouldn't share their passwords either. It's a special secret just for them. Remember that they should only share their password with their teacher, a parent, or a trusted person.

Cybersecurity

Despite the importance of cybersecurity, there is no standard, globally accepted definition for this term.

According to the Cybersecurity and Infrastructure Security Agency (CISA), cybersecurity is defined as....

> ...the art of protecting networks, devices, and data from unauthorized access or criminal use and the practice of ensuring confidentiality, integrity, and availability of information.[24]

Recognizing the fact that young learners soak up the world around them like sponges, use this opportunity to explain and, whenever possible, demonstrate to your students how you approach keeping your digital world and digital assets safe and secure. They will watch you, listen to you, and mimic you. Knowing this provides a platform for discussing the general concept of cybersecurity with them whenever the opportunity presents itself.

Once again, using analogies with young learners is an excellent way to approach the broader discussion of cybersecurity. For example...

Cybersecurity is like a moat that protects a castle (see Figure 2.9). Explain that a moat is like a special trench that goes around a castle to keep it safe. This trench is filled with water, and it helps protect the castle from enemies because they can't easily cross the water to get inside.

Go on to explain that, just like the castle's moat, cybersecurity helps to protect their computers and the information inside them from someone who might try to do something they shouldn't with their computer, tablet, or mobile phone.

Figure 2.9 Castle and protective moat.[25]

Finally reinforce that, as they learned to look both ways before crossing the road safely or not talk to strangers, cybersecurity helps us stay safe in the digital world. It's like having a set of rules and tools to make sure the Internet is a fun and friendly place.

So, when we use computers and tablets to play games or learn new things, cybersecurity helps make sure everything is protected. It's like having a moat around your computer, making sure you have a good time online without any worries!

> If technology is to be used in the classroom, there must also be an emphasis on teaching students safe behavior, and how to protect themselves against inherent technology risks.[26]

Summary

The Condition of Education 2020 report published by the National Center for Education Statistics (NCES) disclosed that in 2018, some 94% of 3- to 18-year-olds had home Internet access: 88% had access through a computer, and 6% had access only through a smartphone, the remaining 6% had no access at all.[27]

Crimes against children and youth and the tactics to ensnare them are becoming more sophisticated. The reason is that children often use devices both inside and outside of school.

As students use technology to support their learning, schools are faced with a growing need to protect student privacy continuously while allowing the appropriate use of data to personalize learning, advance research, and visualize student progress for families and teachers.[28]

The society and world in which students are living and learning continue to become even more digitally connected and dependent; therefore, the concepts and applications of cyber safety and cybersecurity must be incorporated into the academic curriculum and classroom exercises and through ongoing discussions with young learners about the responsible and safe use of technology.

By nurturing cybersecurity awareness among children, they can become responsible digital citizens, equipped with the expertise needed to protect themselves and contribute to a safer online environment. Early exposure paves the way for the next generation of leaders to manage increasing cyber threats effectively.[29]

LESSON PLANS

Kindergarten

Nurturing Digital Citizens

TOC Title: G K Digital Citizenship

Lesson Title: Digital Citizenship/Digital Footprints
Grade Level: Kindergarten
Duration: 25–30 minutes

Objective:

- Students will understand the basic concept of digital footprints.
- Students will understand how to be mindful of their actions when online.

Suggested Materials:

- Large paper or poster board.
- Markers.
- Pictures of footprints (such as the pictures of animal footprints in Chapter. 2).
- Internet safety videos/animations suitable for young children (optional) Suggestions: Follow the Digital Trail, https://www.youtube.com/watch?v=7bRZdUtmH8k (2 minutes, 33 seconds).

  and/or

  What is a digital footprint|Online Safety for Kids|Digital Footprint for Kids|Online Safety https://www.youtube.com/watch?v=tiUYMcPAI84 (1 minute, 20 seconds).

Procedure

Introduction (10 minutes):

a. Whole group: Ask students if they've ever seen footprints in sand or mud and what they think footprints can tell us.
b. Explain that, like footprints on the ground tell a story of where someone has been, our actions on the Internet leave a trail called a "digital footprint."

c. When we go online to play games or watch videos, send messages, or whatever we are doing, we leave a digital footprint that does not go away.
d. Show the video(s) to the class and briefly discuss afterward.

Activity (10–15 minutes):

a. Create a visual aid together. Draw a large footprint on paper or poster board, or have the student outline their footprint. Explain that this is like our digital footprint.
b. Ask students to suggest something that should go on the footprint, e.g., a game they like to play online, a video they enjoy watching, and so on. The teacher writes or draws on the footprint as the students make suggestions.
c. While the class contributes, discuss how each action they take online adds to their digital footprint. Point out that, like in the video, this footprint can be hard to erase.
d. Remind students of the importance of asking parents and trusted adults before doing anything online.

Conclusion (5 minutes):

a. Summarize the lesson and the importance of being safe and responsible online.

Extension Activities:

a. Send a message to parents summarizing what was discussed in class. Encourage them to talk with their child about their online activities.

TOC Title: G K Cyber safety

Lesson Title: Cyber safety

Grade Level: K

Duration: 55 minutes (Consider splitting this lesson plan into two sessions of 35 and 25 minutes each. A suggested break point is after the discussion and before the activity.)

Objective:

- Students will understand the basic concept of cyber safety and the importance of staying safe online.

Suggested Materials:

- Whiteboard or chart paper
- Markers
- Internet safety rules (You may consider using the ABIDE principles in Chapter 2 or create a new acronym.)
- Picture books on cyber safety such as Nettie in Cyberland: Introduce Cyber Security to Your Children by Wendy Goucher, Once Upon a Time Online by David Bedford, or Chicken Clicking by Jeanne Willis
- Drawing materials (crayons, colored pencils, etc.)

Procedure

Introduction (10 minutes):

a. Ask students what they know about computers or the Internet. List their responses on the whiteboard.
b. Explain that today, you'll talk about how to be safe when using computers or tablets, just like being safe when playing outside.

Read Aloud (15 minutes):

a. Use one of the suggested books or another age-appropriate book of the same topic. Pause occasionally to ask questions and discuss the story:
    - What did you notice about what the character did that was safe or unsafe?
    - How did the character feel when something bad happened online?
    - What are some safe things the character could have done instead?

Class Discussion (10 minutes):

a. Use the whiteboard or chart paper to draw simple illustrations of safe and unsafe online behaviors based on the story.
b. Discuss these behaviors with the class:
    - Sharing personal information
    - Clicking on unknown links or pop-ups
    - Talking to strangers online
    - Being kind and respectful in online interactions
c. Ask the children for their thoughts on how to stay safe online.

Suggested break point if splitting into two sessions. Spend a few minutes at the start of Session 2 on a summary of where Session 1 ended.

Activity (15 minutes):

a. Hand out drawing materials, or use the free online coloring page, and ask the children to draw a picture of themselves using the Internet safely.
b. Encourage them to illustrate what safe online behavior looks like based on the discussion.
c. After they finish, have each child describe their drawing and explain how they are being safe online.

Conclusion (5 minutes):

a. Summarize the key points discussed about cyber safety.
b. Remind the children about the importance of asking a trusted adult for help if something online makes them uncomfortable or confused.

Extension Activities:

a. Create a classroom cyber safety poster with drawings or guidelines from the children to display as a reminder.

Grade 1

Nurturing Digital Citizens:

TOC Title: G1 Digital Citizenship

Lesson Title: Digital Citizenship (Strong Passwords)

Grade Level: 1

Duration: 35 minutes

Objective:

- The students will understand the importance of creating strong passwords as part of being good digital citizens.
- Students will demonstrate the ability to create strong passwords.

Suggested Materials:

- Printed worksheets or blank paper.
- Pencils or markers.

Procedure

Introduction (10 minutes):

a. As a group, explain that passwords are like secret codes that help keep our information safe on computers and other devices. (A difficult secret code is stronger than a simple one).
b. Using relatable examples, explain why it's important to have a strong password. Use language and examples from Chapter 2 content.

Activity (20 minutes):

a. Show examples of weak passwords (like "123456" or "password") and explain why they're not safe (too easy for others to guess).
b. Discuss what makes a password strong: Made up of eight or more characters. A mix of uppercase and lowercase letters, numbers, and symbols that they will be able to remember but will be hard for others to guess.

c. Have students create their strong passwords using a worksheet or blank paper.
d. Share ideas/tricks to help remember a password (Chapter 2 examples such as bluestar7 – draw a blue star and the number 7).
e. Encourage creativity, but emphasize that passwords should be kept private and not shared with others, just like a secret code.

Conclusion (5 minutes):

a. Recap what was learned about strong passwords and why they're important.
b. Remind students to never share passwords with anyone except their parents or guardians.

TOC Title: G1 Cyber safety

Lesson Title: Cyber safety

Grade Level: 1

Duration: 25–30 minutes

Objective:

- The students will demonstrate understanding of basic cyber safety rules and practices.

Suggested Materials:

- Whiteboard/markers or a large paper/board
- Colored markers/crayons
- Printed images of online platforms (e.g., a computer, tablet, smartphone). Catalogs or magazines may be used to cut out and collect pictures
- Poster board for a class-created cyber safety poster
- Child-friendly Internet safety videos or rcsources; suggested video: Internet Safety Tips for Kids (https://www.youtube.com/watch?v=qtJNRxMRuPE), or another age-appropriate video on cyber safety.

Procedure

Introduction (5 minutes):

a. Begin with a discussion about technology. (Refer to the chapter as needed.) Ask students what devices they use? What do they like to do online?
b. Explain that while the Internet is fun, there are rules to follow to stay safe, just like when playing games.

Watch Video (Suggested Video is 2:08):

a. Briefly discuss the tips shared in the video.

Activity (15 minutes):

a. Create a class poster titled "Our Cyber-Safety Rules."
b. Discuss and list basic rules together, such as "Always ask a grown-up before using the Internet," "Never share personal information like your name, address, or school online," and "Be kind to others online, just like in real life."
c. Encourage students to draw or write on the poster to illustrate these rules.

Conclusion (5 minutes):

a. Recap the main points of cyber safety using the poster the students created.
b. Encourage students to share the Internet safety tips with their families.

Extension Activity:

a. Send home a simple handout summarizing the key points discussed in class so families can reinforce these safety practices at home.
b. Ask students if they shared the handout with their family and the family response.

Grade 2

Nurturing Digital Citizens:

TOC Title: G2 Digital Citizenship

Lesson Title: Digital Citizenship (Be Kind Online)

Grade Level: 2

Duration: 35 minutes

Objective:

- Students will understand the concept of digital citizenship (being safe, kind, and responsible online).

Suggested Materials:

- A picture book about online safety if available such as The Technology Tail by Julia Cook or another book that addresses the topic at an age-appropriate level.
- Large paper or whiteboard.
- Crayons/markers.
- Poster board.
- Printed images of various devices (computer, tablet, smartphone).
- Internet safety posters or visuals (recommended).

Procedure

Introduction (10 minutes):

a. Start with a class discussion about what the Internet is. Ask students if they have used any devices connected to it.
b. Explain that the Internet is a place where people can find information, play games, and connect with friends, but it is important to know how to be safe, responsible, and kind when online.

Read Aloud (10 minutes):

a. If available, read a picture book that addresses the concept of online safety, focusing on appropriate, kind, and responsible behavior.

b. As you read and afterward, ask students questions related to the story to gauge understanding and reinforce key points.

Class Discussion/Activity (10 minutes):

a. Use a large piece of paper or poster to write out a list of rules for being safe and respectful online. e.g., never post something that might hurt someone, never give out personal information, and never talk to strangers. Take suggestions from students to add to the class poster.
b. Discuss these rules and why they are important. Refer to the content in Chapter 2.

Conclusion (5 minutes):

a. Display the poster in the classroom.
b. Summarize what it means to be a good digital citizen, i.e., being safe, kind, and asking for help when needed.

Extension Activities:

a. Send a note home to parents summarizing what was learned in class about digital citizenship.

TOC Title: G2 Cyber safety

Lesson Title: Cyber safety on the Internet
Grade Level: 2
Duration: 45 minutes

Objective:

- The students will understand the basics of online safety and know the rules of cyber safety.

Suggested Materials:

- Whiteboard and markers, or a projector for visuals.
- Internet safety posters such as printouts of the ABIDE principles in Chapter 2.
- Paper and coloring materials.
- Educational videos or interactive games about cyber safety (age-appropriate) Suggestions: Being Safe on the Internet, https://www.youtube.com/watch?v=HxySrSbSY7o by AMAZE.org
  or
- Online Privacy for Kids – Internet Safety and Security for Kids
  https://www.youtube.com/watch?v=yiKeLOKc1tw By Smile and Learn

Procedure

Introduction (5 minutes):

a. Class discussion: ask students about their favorite things to do online. Watch videos? Play games?
b. Explain that there are rules for staying safe when using computers, tablets, or phones, just like we have rules for staying safe in the real world.

Discussion on Online Safety Rules (10 minutes):

a. Discuss basic online safety rules in age-appropriate terms:
b. Never share personal information (name, address, or phone number) online.

c. Always ask a grown-up before clicking on links or downloading anything.
d Be kind and respectful to others online, just like in real life.
e. Use relatable examples to illustrate each rule.

Video and Discussion (10 minutes):

a. Show an age-appropriate educational video (some options suggested above) that reinforces cyber safety concepts learned in class.
b. Discuss important topics from the video.

Activity (15 minutes):

a. Cyber Safety Poster Creation
    - Divide the class into small groups.
    - Provide each group with materials to create a cyber safety poster illustrating one important rule they learned. Consider assigning topics to groups to ensure everything is covered.
    - Encourage creativity and clarity in posters.

Conclusion (5 minutes):

a. Summarize the main cyber safety rules discussed during the lesson.

Homework (optional):

a. Encourage students to share what they learned about cyber safety with their families.

Notes

1 Nevens, T. M., "Fast Lines at Digital High," The McKinsey Quarterly, Winter 2001, pp. 167–177. Gale Academic OneFile, https://link.gale.com/apps/doc/A72524631/AONE?u=anon~3971fa9e&sid=googleScholar&xid=3be7d694, retrieved November 25, 2023.

2 The overview of the pervasiveness of technology was developed based on a review and summarization of multiple documents and supported in part, through references from the following sources: (a) Haleem, A., Javaid, A., Qadri, M., & Suman, R., May 23, 2022, "Understanding the Role of Digital Technologies in Education: A Review," Sustainable Operations and Computers, pp. 275–285, Published by Elsevier B.V. on behalf of KeAi Communications Co., Ltd. This is an open-access article under the CC BY license (http://creativecommons.org/licenses/by/4.0/), (b) Lucke, U., & Rensing, C., October 2014, "A Survey on Pervasive Education," *Pervasive and Mobile Computing*, (14), 3–16, ISSN 1574-1192, (c) Cladis, A., September 2020, "A Shifting Paradigm: An Evaluation of the Pervasive Effects of Digital Technologies on Language Expression, Creativity, Critical Thinking, Political Discourse, and Interactive Processes of Human Communications," *E-Learning and Digital Media*, 17(5), 341–364, https://eric.ed.gov/?id=EJ1260987, and (d) Shubina, I., & Kulakli, A., 2019, Pervasive Learning and Technology Usage for Creativity Development in Education, *International Journal of Emerging Technologies in Learning (iJET)*, 14, 95. https://doi.org/10.3991/ijet.v14i01.9067, retrieved November 26, 2023.

3 Auxier, B., Anderson, M., Perrin A., & Turner, E., July 28, 2020, "Parenting Children in the Age of Screens," Pew Research Center, www.pewresearch.org/internet/2020/07/28/parenting-children-in-the-age-of-screens, and www.pewresearch.org/internet/2020/07/28/childrens-engagement-with-digital-devices-screen-time, retrieved December 21, 2023.

4 EdWeek Research Center, 2022, "Technology in Teaching and Learning," https://epe.brightspotcdn.com/8d/b6/49769ee54be9af7ed5287b6b2a0a/technology-in-teaching-and-learning-research-spotlight-4.13.22_Sponsored.pdf, retrieved December 21, 2023.

5 Hanover Research, January 25, 2023, "2023 Trends in K–12 Education," www.hanoverresearch.com/reports-and-briefs/2023-trends-in-k-12-education/?org=k-12-education, retrieved November 29, 2023.

6 Thomas, S., et al., October 2016, "Policy Brief on Early Learning and Use of Technology," U.S. Department of Education, Office of Educational Technology, Washington, D.C., http://tech.ed.gov/earlylearning.

7 Karppinen, I., Nurse, J. R. C., & Varughese, J., 2023, "Oh Behave! The Annual Cybersecurity Attitudes and Behaviors Report 2023," The National Cybersecurity Alliance and CybSafe, www.cybsafe.com/whitepapers/cybersecurity-attitudes-and-behaviors-report, retrieved November 29, 2023.

8 Snyder, S., 2016, "Teachers' Perceptions of Digital Citizenship Development in Middle School Students Using Social Media and Global Collaborative Projects," Walden Dissertations and Doctoral Studies, https://scholarworks.waldenu.edu/dissertations/2504, retrieved December 6, 2023.

9 U.S. Department of Education, Office of Educational Technology, January 2017, "Reimagining the Role of Technology in Education: 2017 National Education Technology Plan Update," Department of Education, Washington, D.C., http://tech.ed.gov, retrieved December 4, 2023.

10 Ibid.

11 Rachel's Challenge, 2023, "Digital Citizenship for Students," https://rachelschallenge.org/get-info/digital-citizenship, retrieved December 4, 2023.

12 Figure 2.1 drawn by the authors based upon the International Society for Technology in Education (ISTE) standards, https://iste.org/standards, retrieved December 6, 2023.

13 Figure 2.1 drawn by the authors based on the International Society for Technology in Education (ISTE) standards.

14 Centre for the Protection of National Infrastructure, 2016, "My Digital Footprint: A Guide to Digital Footprint Discovery and Management," National Protective Security Authority, www.npsa.gov.uk/resources/my-digital-footprint-brief-guide, retrieved December 12, 2023.

15 Ibid.

16 February 18, 2019, A Lion's Paw Prints in The Sand, Photo by Yassine Khalfalli on Unsplash at https://unsplash.com/@yassine_khalfalli?utm_content=creditCopyText&utm_medium=referral&utm_source=unsplash (Yassine Khalfalli) on https://unsplash.com/photos/paw-print-on-sand-adsQ9d_Cx0c?utm_content=creditCopyText&utm_medium=referral&utm_source=unsplash (Unsplash), retrieved December 12, 2023.

17 June 20, 2018, Animal Tracks in the Snow, Photo by Enrico Mantegazza on Unsplash, at https://unsplash.com/@limpido?utm_content=creditCopyText&utm_medium=referral&utm_source=unsplash (Enrico Mantegazza) on https://unsplash.com/photos/snow-paw-prints-SIgxkS-ZPm8?utm_content=creditCopyText&utm_medium=referral&utm_source=unsplash (Unsplash), retrieved December 12, 2023.

18 OpenAI's DALL-E, December 29, 2023, "Puppy Leaving Scent Marks," generated using DALL·E, via author's OpenAI user license, produced December 29, 2023.

19 August 8, 2023, Bird Feather, Photo by Josie Weiss, on Unsplash, at https://unsplash.com/@scarlettweiss?utm_content=creditCopyText&utm_medium=referral&utm_source=unsplash (Josie Weiss) on https://unsplash.com/photos/a-white-feather-sitting-on-top-of-a-lush-g

reen-field-BP0nKvxXGDs?utm_content=creditCopyText&utm_medium=referral&utm_source=unsplash (Unsplash), retrieved December 12, 2023.

20 November 13, 2022, Photo Spider on Leaf, by B_Man, on Unsplash, at https://unsplash.com/@b_man?utm_content=creditCopyText&utm_medium=referral&utm_source=unsplash (Blind Man) on https://unsplash.com/photos/a-moth-on-a-leaf-V6YiPxlY64I?utm_content=creditCopyText&utm_medium=referral&utm_source=unsplash (Unsplash), retrieved December 12, 2023.

21 Pau, K. N., Lee, V. W. Q., Ooi, S. Y., & Pang, Y. H., March 13, 2023, "The Development of a Data Collection and Browser Fingerprinting System," *Sensors (Basel, Switzerland)*, 23(6), 3087. https://doi.org/10.3390/s23063087, www.ncbi.nlm.nih.gov/pmc/articles/PMC10057587/, retrieved December 13, 2023.

22 Image of Magical Paintbrushes," created by OpenAI's DALL-E, accessed January 20, 2024, generated using ChatGPT with Image Generator Tool.

23 May 16, 2019, Toys, Photo by Huy Hung Trinh, in Unsplash, at https://unsplash.com/@trinhhuyhung?utm_content=creditCopyText&utm_medium=referral&utm_source=unsplash (Huy Hung Trinh) on https://unsplash.com/photos/assorted-colored-toys-on-table-zoyBqT7ytLU?utm_content=creditCopyText&utm_medium=referral&utm_source=unsplash (Unsplash), retrieved December 14, 2023.

24 Cybersecurity and Infrastructure Security Agency (CISA), February 1, 2021, "What Is Cybersecurity?," www.cisa.gov/news-events/news/what-cybersecurity, retrieved December 15, 2023.

25 "Castle with Protective Moat," generated by OpenAI's DALL-E, accessed January 20, 2024, created using ChatGPT with Image Generator Tool.

26 Safe Search Kids, (n.d.), "The Risks of Technology as an Aid in K-12 Education," www.safesearchkids.com/the-risks-of-technology-as-an-aid-in-k-12-education, retrieved December 15, 2023.

27 Hussar, B., NCES, Zhang, J., Hein, S., Wang, K., Roberts, A., Cui, J., Smith, M., AIR, Bullock Mann, F., Barmer, A., Dilig, R., & RTI, May 19, 2020, "The Condition of Education 2020," National Center for Education Statistics (NCES), https://nces.ed.gov/pubsearch/pubsinfo.asp?pubid=2020144, retrieved December 15, 2023.

28 U.S. Department of Education, (n.d.), National Education Technology Plan (NETP), https://tech.ed.gov/netp/introduction, retrieved December 15, 2023.

29 Safe Search Kids, (n.d.), "Teaching Kids about Cybersecurity: Engaging Methods for Young Minds," https://www.safesearchkids.com/teaching-kids-about-cybersecurity-engaging-methods-for-young-minds/#:~:text=By%20nurturing%20cybersecurity%20awareness%20among,manage%20increasing%20cyber%20threats%20effectively, retrieved December 15, 2023.

3
STRANGERS
Safe/Unsafe People and the Tricky Person

Introduction

As we move from discussing character education and nurturing digital citizens, we proceed to another equally important area, yet one that may be difficult to explain to younger students.

Who Is a Tricky Person?[1]

Previously, danger was an unknown bad guy who presented *him*self in person and was named stranger danger.

The threat has moved from strictly an in-person encounter to include an online encounter. There is often a hybrid version where the online meeting is nurtured and is subsequently used to gain trust to facilitate an in-person meeting. This hybrid created the term tricky person.

At one time, it was considered that danger came in the form of a stranger – "Don't talk to strangers." "If you don't know someone, do not talk to or go with them." All sound advice, but we now know that danger may also be encountered in the form of a person, or someone known to the child and often known to the family. Someone may be a stranger, but not a dangerous person, and is a helpful and needed person, like a police officer or a firefighter.

Consider the confusion created by telling a child not to talk to strangers and then they are lost, or siblings are together at home and there is an emergency. The child needs to reach out for help. To the young mind, these helpers are strangers. The idea of a tricky person conveys to the child that there is deception in trust. In tandem with the deception and trust concepts, the term safe or unsafe person is also

DOI: 10.1201/9781003465928-3

Figure 3.1 Safe person (a) vs. tricky person (b).[3]

used. Using both tricky and unsafe people speaks to actions and less about the relationship involved (see Figure 3.1).

According to the Centers for Disease Control and Prevention, someone known and trusted by the child or the child's family members perpetrates 91% of child sexual abuse.[2]

Who Is and Is Not a Tricky or Unsafe Person?

How to talk with children at the earliest ages about a tricky person and deception, and who is a safe or unsafe person?

The website Lifehacker's article on teaching children about safe people with TikTok creator and parenting expert, Jessica Martini, suggests sharing the following parameters with kids to identify safe adults in their lives, rather than creating a traditional "list" by names or titles. These parameters are good to practice when talking about safety with children.[4]

- We can begin by telling our kids: "A safe adult is someone who makes you feel happy and safe," Martini explains. "When you go near them, you don't feel nervous, scared, or have an icky feeling in your tummy. They make you feel loved and comfortable."
- Secondly, a safe adult (or adolescent) "will never, ever ask you to keep a secret," and if they do, instruct your child to share the secret with you right away. She also advises not to

teach kids about "good vs. bad secrets" (ones that make them feel happy vs. bad) because sometimes abusers will use "good secrets to get their foot in the door." E.g., "Here's a cookie, don't tell your mom" or "I know you broke that, but I won't tell anyone, it's our little secret." Getting kids in the habit of keeping "good secrets" makes it easier for them to keep bad secrets in the future. Instead, Martini notes, we can teach them the difference between secrets and "happy surprises"—things that everyone will find out soon and everyone will feel happy about, such as a surprise party.

- A safe adult will *always believe you* when you tell them something important. Martini points out that sometimes our kids will "tell us without telling us" when they don't have the words or emotional maturity to convey it verbally. They may act out the traumatic situation in their play, experience headaches, stomach aches (sic), or other signs of physical illness (especially when they're going to see a certain person), become withdrawn, or experience a loss of appetite. They may also act distressed when going somewhere they previously enjoyed going – exhibiting behaviors, such as screaming, crying, clinging, refusing to move, or saying, "They're the worst, I never want to go there."

Kids First, Inc., started in 1991 and serves up to 250 children per year. It provides the following list of subjects to discuss with children about body safety and the safe/unsafe person concept.

- Talk about "safe" and "unsafe" touching rather than "good" or "bad" touching. This removes guilt from the child and keeps them from having to make a moral distinction about what is and is not appropriate.
- Use age-appropriate wording. You can discuss body safety without discussing sexuality. Teach young children that no one should touch them in any area that their bathing suit covers, and that they should never touch anyone else in these areas or see pictures or movies that show those areas.

- Teach the difference between healthy and unhealthy secrets. An example is that a surprise party is an okay secret to keep because it will make people happy and will be told at the right time. Secret touching is not okay, nor is keeping any permanent secrets from parents or caregivers.
- Have your child identify five people that they could/would talk to if someone was touching them in an unsafe way. Children are often afraid to tell their parents out of fear of punishment (or because of a threat made by a perpetrator), so your child needs to know they can seek out other trusted adults to confide in. Instruct your child that they should keep telling you until someone helps them.
- Teach children proper names for body parts so that if they disclose inappropriate touching, it will be clear what they are talking about.
- Revisit this safety talk often. Children learn through repetition. How many times do you remind children to look both ways before crossing the street? (Figures 3.2–3.5).[5]

Planet Puberty, a digital resource suite by Family Planning NSW, aims to provide parents and carers of children with intellectual disability and/or autism spectrum disorder with the latest information, strategies, and resources for supporting their child through puberty. Planet Puberty's focus is on children with intellectual disabilities and autism spectrum disorders, the list of rules for tricky people may be used in that environment and is also applicable to all children.[10]

The following are some rules for children with intellectual disability and autism spectrum disorder about 'tricky people' (summarized from the Safely Ever after Program).

1. People must ask your permission before touching you in any way. This includes hugs and kisses. Your child does not need to apologize or have a reason for not wanting to hug or kiss someone.
2. Everybody has private body parts that must be covered when you are in public.

Child Safety Disclosure What to Say

- Thank you for telling me.
- I believe you.
- I will support you.
- Everyone has the right to feel safe.
- There is nothing so bad that you can't tell someone.

Figure 3.2 What to say.[6]

Child Safety Disclosure What to Say

- I am sorry that happened.
- Thank you for trusting me.
- It is not your fault.
- You are brave telling me.
- You did the right thing telling me.

Figure 3.3 What to say (continued).[7]

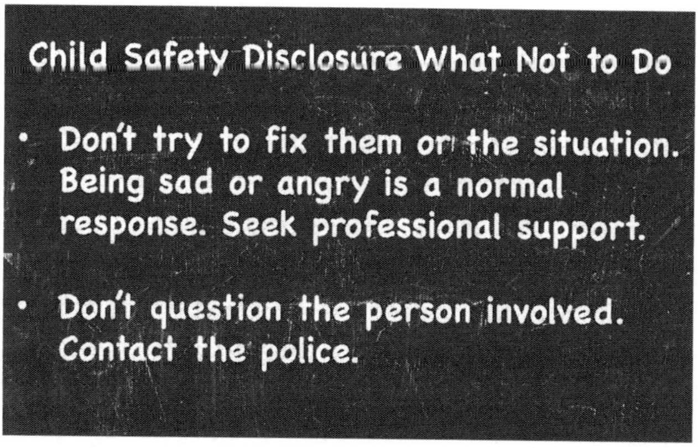

Child Safety Disclosure What Not to Do

- Don't try to fix them or the situation. Being sad or angry is a normal response. Seek professional support.
- Don't question the person involved. Contact the police.

Figure 3.4 What not to do.[8]

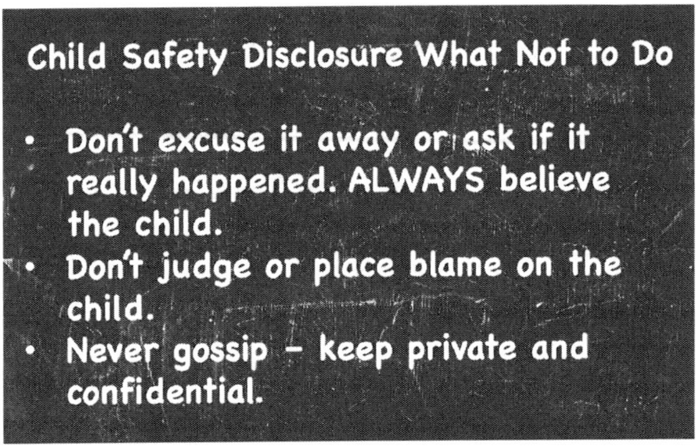

Child Safety Disclosure What Not to Do

- Don't excuse it away or ask if it really happened. ALWAYS believe the child.
- Don't judge or place blame on the child.
- Never gossip – keep private and confidential.

Figure 3.5 What not to do (continued).[9]

3. Make sure your child knows their full name along with the full names of their parents/caregivers.
4. Your child should never go anywhere with someone they don't know or take anything from someone they don't know. Things like ordering takeaway or asking for something at the shops are different because you are requesting help from someone.
5. Your child should always check with their parents/caregivers before:
   - changing plans without prior notice
   - getting into a car (even if the driver is someone they know)
   - accepting gifts (gifts should never be a secret)
6. Your child does not have to be polite if they feel scared or uncomfortable.
7. Your child is allowed to say or sign NO to adults and other children.
8. No, Go, Tell: Teach your child to say or sign no, and to then go and tell their safe people.
9. Your child should never be asked to keep a secret. Secrets can make us feel scared and uneasy. If there is information your child must keep to themselves, like a party or a gift for another person, reframe this as a SURPRISE! No adult should ever ask your child to keep a secret.

Remember that safe touch is good! Don't forget to give your child examples of positive, safe touch. For example, high fives with friends or a hug from a safe adult that makes you both feel good.

A good exercise may be to have students identify who their safe people are. Create a classroom list of safe people, encouraging students to share their ideas of safe people. Be sure to continue supporting good and positive examples.

Teachers can be a safe, trusted adult for children.

Darkness to Light, an organization with the mission to end child sexual abuse, provides methods for you to be a safe adult – so that a child can trust you to be there for them when they need help.

- Set and Maintain Clear, Protective Boundaries: This is a crucial step because boundaries help us to respect each other and feel respected. A child who knows that you will respect (and even protect!) their boundaries will have an easier time trusting you to take care of them.
- Develop Protective Bonds: Really listen when the child talks to you. Show them that you're interested in their opinion; involve them in conversations; and show them that their input is valuable.
- Talk Openly and Honestly about Child Sexual Abuse: When we talk to children in age-appropriate ways about our bodies, sex, and boundaries, children understand what healthy relationships look like. It also teaches them that they have the right to say "no."[11]

The National Center for Missing and Exploited Children® suggests using the following language when talking to your child about abduction prevention:

<u>DON'T SAY</u>

Never talk to strangers.

<u>DO SAY</u>

You should not approach just anyone. If you need help, look for a uniformed police officer, a store clerk with a nametag, or a parent with children.

DON'T SAY

Stay away from people you don't know.

DO SAY

It is important to get my permission before going anywhere with anyone.

DON'T SAY

You can tell someone is bad just by looking at them.

DO SAY

Pay attention to what people do. Tell me right away if anyone asks you to keep a secret, makes you feel uncomfortable, or tries to get you to go somewhere with them.[12]

These tips and the language may be adapted and used in various discussions about tricky and safe/unsafe people.

What's a Family Safe Word?

One way to ensure that someone is not dangerous, tricky or unsafe is to create a family safe word. This is a word or a couple of words that are known to only the immediate family. It is not shared with other relatives, friends, or neighbors. It is easy enough for children, even at the youngest of ages, to remember.

How and When to Use the Family Safe Word

The safe word may be used by the child in any environment where a child feels unsafe or uncomfortable with their surroundings. It makes them feel that they have control in many situations. An example provided by The Pragmatic Parent places the child at a friend's BBQ where the child is with an adult who is making inappropriate conversations and requests of your child. Your child can come to you and tell you the safe word regardless of whether you are alone or standing around in a group with your friends – you will know exactly what they are disclosing and can respond immediately.[13]

A safe word can be used anywhere – you may even create a classroom-safe word. This allows you and the child to have a code for all situations. The importance of using the safe word must be strongly conveyed – think of the child who cried wolf one too many times. Examples of the safe word may be used at a school sporting event, after-school activities, or any school-sponsored event where the child feels uncomfortable around an adult or even another student.

Summary

By raising awareness among children, they can learn to become responsible for understanding that they can say "No" to an adult if they are uncomfortable or it feels "icky" and that they have the tools to identify who the helpers are and who may be playing a trick on them. It is not too early to create a sense of ownership of the body and personal safety.

Chapter 4 presents the emerging topic of "Exploring Cyber Safety through Stewardship."

LESSON PLANS

Kindergarten

Strangers: Safe/Unsafe People and the Tricky Person

TOC Title: G-K Tricky Person

Lesson Title: Tricky Person

Grade Level: Kindergarten

Duration: 65 minutes (It is recommended to break this into three lessons. Suggested breakpoints are included.)

Objective:

- The students will be able to define and identify safe people and tricky people.
- Students will know how to seek help when they encounter a tricky person.

Suggested Materials:

- Pictures representing family members, strangers, community helpers, online avatars, etc.).
- Whiteboard or large flip chart and markers.
- If available: Picture books or videos addressing this topic such as Protect Yourself Rules – A Friendly Stranger and The Berenstain Bears Learn about Strangers (also a book).
- Other suggestions that should be previewed by teachers/administration/parents due to sensitive topics: Some Secrets Should Never Be Kept by Jayneen Sanders (a storybook – can be found as a read aloud on YouTube) and My Body Belongs to Me by Jill Starishevsky (a picture book – there are read-aloud versions on YouTube).

Procedure

Introduction (15 minutes):

a. Whole group: discuss feelings, i.e., “What does it feel like when you are happy or sad?”

b. Introduce the terms Tricky Person and Safe Person. Explain that safe people help us feel safe and loved. Tricky people might make us feel uncomfortable, scared, confused, and icky.
c. Use pictures to help identify safe and tricky people. Encourage discussion about why students think someone may be tricky or safe.
d. Include in the discussion that it may be harder to tell online if someone is safe or not. Encourage students to pay attention to how their online person makes them feel. Tell students, "Someone may seem nice at first, but they may begin saying, asking or doing things that make you feel uncomfortable. That is a sign that they are not safe."

Recommended end of Part 1 if doing more than one lesson.

Part 2:

Read Aloud or Watch the Video (10 minutes) [Optional if books/videos are not available go to class discussion. Students will probably need more guidance regarding the behavior of tricky and safe people].

Class Discussion and Activity (25 minutes)

a. Start by briefly reviewing the story or video
   - Discuss the characters' actions and feelings.
   - Role-playing activity: assign roles where some students exhibit safe person behaviors and others exhibit tricky person behaviors. Let students interact and identify other students as safe or tricky.
     - Encourage students to decide what they should do next if they think the person they are with is a tricky person (e.g., run away, tell a trusted adult, call for help).
   - Make a chart on the whiteboard or flipchart listing the traits of safe people and the traits of tricky people.

Suggested end of Part 2 if doing three lessons.

Part 3

Conclusion (15 minutes):

a. Review the characteristics of safe and tricky people. Highlight that tricky people do not always seem scary or mean in person or online.
b. Go over these safety rules:
    - Never go anywhere with any person you are not expecting to take you (i.e., pick you up from school or a sports activity, etc.), even if they are not a stranger.
    - Always check with a trusted adult before going somewhere, including when you are online.
    - Never share personal information unless a trusted adult tells you it is safe.
c. Encourage students to identify safe adults they can talk to if they feel uncomfortable or scared.

Homework:

a. Provide a handout summarizing the lesson for students to share and discuss with their families.

Extension Activities:

a. Revisit this topic throughout the year.
b. Suggested activities: Coloring or drawing activities depicting safe and tricky people, role-playing activities where students practice saying "no" to tricky people and getting help.

Grade 1

Strangers: Safe/Unsafe People and the Tricky Person

TOC Topic: Strangers: Safe/Unsafe People and the Tricky Person

Lesson Title: Class Share ... Just Shout "NO!"

Grade Level: 1

Duration: 40 minutes

Objective:

- Introduce the concept of a tricky person so that the child knows this is not a person to be trusted.
- Provide students with a tool to respond when encountering a tricky and unsafc person.

Suggested Materials:

- A whiteboard or chart paper with markers.
- Drawing matcrials (crayons, markers, colored pencils).
- Suggested book: Shout NO!: A Child's Rhyme about Tricky People...And What to Do by Sarah Ernst, or another book that addresses unsafe and tricky people at an age-appropriate level.

Procedure

Class Discussion (15 minutes):

a. Brainstorm together who might be considered a tricky person and why that makes them unsafe (examples may include a stranger who is not a helper, like a police officer, a firefighter, or the helper if they need to call 911).
b. Give the class realistic scenarios with choices to help them understand a safe or unsafe person.

Activity (15 minutes):

a. Read the suggested or similar book and discuss (or act out) how to respond to an unsafe/tricky person.
b. Encourage students to feel comfortable saying firmly, yet politely, tell a grown-up "No" and to walk away.

Conclusion (10 minutes):

a. Discuss the importance of knowing who to trust.
b. Discuss why it is important to know who a safe person is, and not an unsafe or tricky person.

Grade 2

Strangers: Safe/Unsafe People and the Tricky Person

TOC Topic: Strangers: Safe/Unsafe People and the Tricky Person

Lesson Title: Not Everyone Is Nice

Grade Level: 2

Duration: 40 minutes

Objective:

- Introduce the concept of a tricky person so that the child knows this is not a person to be trusted, even though they may seem nice.
- Provide students with a tool to respond when encountering a tricky and unsafe person.

Suggested Materials:

- A whiteboard or chart paper with markers.
- Drawing materials (crayons, markers, colored pencils).
- Pictures from nature of pretty but lethal plants and wildlife.
  - Cute But Deadly: The 10 Most Vicious Animals That Look Cute! – A-Z Animals (a-z-animals.com)
  - Beautiful but Lethal Plants Found in the U.S. (treehugger.com)
- Suggested book: Not Everyone Is Nice by Frederick Alimonti, or another book that addresses verbal bullying at an age-appropriate level.

Class Discussion (15 minutes):

a. Talk to students about what to do if a stranger approaches them, reminding them that looking or acting nice doesn't mean a person is safe or is not trying to trick them into thinking they are safe.
b. Discuss with students how they feel when being tricked.
c. As you demonstrate empathy, you also convey that safety is more important than feeling embarrassed or other emotions.

Activity (15 minutes):

a. Read the suggested, or similar book, and discuss how to respond to an unsafe/tricky person.
b. The suggested book uses examples from nature of pretty but lethal plants and fish to show that animals and people who look or seem nice may be very dangerous. You may incorporate this idea with pictures of such plants, fish or wildlife.

Conclusion (10 minutes):

a. Discuss the importance of knowing who to trust.
b. Discuss why it is important to know who an unsafe or tricky person is and not to be scared to leave and tell someone about them.

Notes

1 Rogers-Nelson, K, January 18, 2018, "'Stranger Danger' Is Over – Here's What Parents Are Teaching Their Kids Instead," https://www.sheknows.com/parenting/articles/1137790/stranger-danger-is-over-tricky-people/, retrieved February 22, 2023.
2 Centers for Disease Control and Prevention, (n.d.), Fast Facts: Preventing Child Abuse, Fast Facts: Preventing Child Sexual Abuse|Violence Prevention|Injury Center|CDC, retrieved January 16, 2024.
3 OpenAI, 2024, "Image of a Trusted Or Safe Person and an Unsafe or Tricky Person," Digital Image. DALL·E. This image was generated upon the author's request via conversation with the AI on January 3, 2024.
4 Showfety, Sarah, July 20, 2022, "The Best Way to Teach Your Kids to Recognize a Safe Adult," https://lifehacker.com/the-best-way-to-teach-your-kids-to-recognize-a-safe-adu-1849194970, retrieved November 27, 2023.
5 Kids First, Inc., (n.d.), https://www.kidsfirstinc.org/how-to-talk-to-young-children-about-body-safety/, retrieved November 27, 2023.
6 Figure 3.2 adapted by authors from The Gentle Counsellor, https://thegentlecounsellor.com/talking-to-your-child-about-safety-strangers-tricky-people/#:~:text=Tip%203%3A%20Teach%20your%20child,someone%20known%20to%20the%20child.
7 Figure 3.3 adapted by authors from The Gentle Counsellor, https://thegentlecounsellor.com/talking-to-your-child-about-safety-strangers-tricky-people/#:~:text=Tip%203%3A%20Teach%20your%20child,someone%20known%20to%20the%20child.
8 Figure 3.4 adapted by authors from The Gentle Counsellor, https://thegentlecounsellor.com/talking-to-your-child-about-safety-strangers-tricky-people/#:~:text=Tip%203%3A%20Teach%20your%20child,someone%20known%20to%20the%20child.
9 Figure 3.5 adapted by authors from The Gentle Counsellor, https://thegentlecounsellor.com/talking-to-your-child-about-safety-strangers-tricky-people/#:~:text=Tip%203%3A%20Teach%20your%20child,someone%20known%20to%20the%20child.
10 Planet Puberty, (n.d.), Identifying Safe People - Planet Puberty, retrieved January 16, 2024.
11 Darkness to Light, April 3, 2020, "How to Be a Safe Adult", https://www.d2l.org/how-to-be-a-safe-adult/, retrieved November 27, 2023.
12 (n.a.), (n.d.), "Rethinking Stranger Danger", Kids Smartz, The National Center for Missing and Exploited Children®, https://www.missingkids.org/education/kidsmartz, retrieved December 1, 2023.
13 The Pragmatic Parent, (n.d.), "Why Every Family Needs a Safe Word," https://www.thepragmaticparent.com/safeword/#:~:text=A%20safe%20word%20is%20a,as%20puppy%2C%20soccer%20or%20milk, retrieved November 27, 2023.

4

Exploring Cyber Safety through Stewardship

Introduction

What has been discussed in previous chapters and learned through in- and out-of-class activities may now culminate in focusing the young learner's attention on acting responsibly when engaging in online activities, on acting safely with their and other people's possessions, with other people, and when interacting with the environment in which they live, learn and play.

Stewardship: Definition and Importance at This Age Level

Stewardship. Why is this important at this age level? First, let us start by defining stewardship. Merriam-Webster defines stewardship as:

> The conducting, supervising, or management of something ...especially the careful and responsible management of something entrusted to one's care.

Our discussion in Chapter 4 focuses on the second half of this definition and streamlines it to state that stewardship is vigilantly taking care of something you have.

Like not bullying others, here the analogy is used to teach children to not bully and be disrespectful to our planet. Building the foundation for stewardship early will create a sense of responsibility in young learners.... the eventual caretakers of our planet Earth.

How Do We Foster Stewardship among Children?

Because science and engineering practices are often underemphasized in K-12 science education, The National Science and Teaching

DOI. 10.1201/9781003465928-4

Association's (NSTA) Environmental Education Professional Development Institute (EEPDI) emphasized pathways to integrating next-generation practices with environmental stewardship for informed, responsible action on behalf of the environment and future generations.[1]

The EEPDI Focus emphasizes the integration of *Next Generation Science Standards* (NGSS) practices through student-centered stewardship projects, which offer an excellent means to get students involved in science, increase their critical thinking, and motivate action on environmental issues that are meaningful to the students. One goal was to engage students in "real" local environmental issues by getting them outdoors working in collaborative groups.

NSTA concludes that you too, like the teachers involved in the Environmental Education Professional Development Institute, can incorporate the *NGSS* Science Practices into your student-driven stewardship projects. These activities can be adapted to student abilities. The project teachers found it helpful to also seek cross-disciplinary (e.g., math teachers) collaboration at their school as well as recruit community expertise outside of their schools.

Science and engineering practices were integrated as students asked questions, planned and carried out their investigation, analyzed and interpreted data, and communicated information. Since repeated exposure is required over many years to deeply understand the disciplinary core ideas and crosscutting concepts (CCCs) and to build students' proficiency with the practices, it is proposed that repeated, varied stewardship projects provide an excellent opportunity for three-dimensional learning.

According to Lesley University,[2] giving children opportunities to be connected to the natural world and ensuring children become ecologically literate citizens who have a sense of ownership and stewardship of the Earth. Environmental educators in particular have been instrumental in creating awareness, programs, and opportunities for children of all ages to connect with the natural world, knowing that our future depends on ecologically literate citizens who will have a sense of ownership and stewardship of the Earth. A No Child Left Inside movement has spawned as a result.

Here are seven specific suggestions from Lesley University's science faculty, which you may use or adapt as age-appropriate.

1. Allow children unfettered time in the natural world.

   This means not organized sports or adult-directed activities, but lengthy time to explore, play, and invent. It might take the form of building a fort from found objects, damming a stream, or collecting natural objects like shells, rocks, or acorns. These moments in nature are recalled later, as adults, in vivid descriptions.
2. Be a mentor.

   Rachel Carson, in her 1956 article entitled "A Sense of Wonder," implored adults to find one child to mentor in the workings of the natural world. Whether parent, grandparent, or other relative or friend, most adult conservationists can point to those people in their lives who had significant influence on them.

   Create opportunities for children to have experiences with the more-than-human world. Volunteering at a wildlife rehabilitation center, snorkeling on a vacation, or simply walking in nature and having surprise contact with local species are all important ways children can come in contact with the species that share their home. Speak of them as friends and talk about how they are connected to humans.
3. Study the local bioregion.

   Guide children to understand the area they live in, for instance, where their water comes from and where it goes once it leaves their home or school; what plants are native or non-native, wild or cultivated; what animals share their home with them; how people make a living from the Earth's resources; and what natural wonders—ponds or streams, marshes, hills, and so on—are nearby.
4. Engage children in real-life actions.

   Whether it is planting trees, creating a garden, pulling invasive species, or picking up garbage—begin the stewardship mindset.

5. Enroll your child in a real outdoor camping program.

   In addition to a physical fitness camp, technology camp, or sports camp, find a camp where children sing by the campfire, sleep out under the stars, learn to make bows and arrows, learn to steer a canoe, or learn to use a bow drill to make a fire.
6. Get to know your state or national parks.

   If you are lucky enough to live near a protected area, visit. These parks are protected for a reason and offer wonderful opportunities for hiking, exploring, and experiencing the sounds, smells, and excitement of unfamiliar natural environments.

Earth.org reinforces connecting students with nature and the outdoors. "Parents and teachers can help students understand their role as environmental stewards by encouraging student outdoor learning programs and supporting young folks who engage in student activism."[3] If your school or nearby park has natural playgrounds, like those built from sustainable materials and found objects, Earth.org indicates this is a perfect place and opportunity to discuss environmental protection and the importance of stewardship over the Earth's resources.

What Does Stewardship Mean for the Teacher's Role?

In the article "The Importance of Stewardship in Leadership" from the Graduate Programs for Educators, the concept of the teacher as a wise steward of leadership in the classroom is examined.[4] Burress believes these are the expected areas of stewardship when serving as an educator and leader. Each of these areas is an important factor in becoming an effective steward of educational leadership.

Stewardship of Students

First and foremost, the most important thing we are entrusted with is our students. What does it mean to be a steward for students? Our communities trust us with their students for 180 days a year. However, more importantly, parents trust us with their children! We need to always ask ourselves the following daily questions:

1. Is what we are doing in the best interest of students?
2. Would you entrust your child to our instructors?
3. Are we making an impact on these children's lives?

As educators, we should never take the responsibility of stewardship of other people's children lightly. Education is a calling; being a wise steward of students is a responsibility that is accompanied by passion, enthusiasm, and joy.

Stewardship of Influence

Educators have a remarkable amount of influence over their students, and educational leaders have the same influence over their staff. We are entrusted to use that influence in such a way that our instructional teams can help our students become productive members of society. We have all heard stories of educators telling students that they are not college material or shouldn't do something because they don't have the right skill set.

Words matter, and misusing words is a poor steward of influence. Look to these daily questions to ensure you are stewarding your influence well:

1. Is my influence helping instructors more effective?
2. Would I be motivated by my words?
3. How can my influence help shape my staff in a way that builds our school culture?

We need to use our influence to lift our teachers up so that they can do the same for our students. Consciously choose to lead staff down a path of success. Look to the positives – not the negatives – of your team and students. Stewardship of influence is a powerful tool, so use it wisely and lead staff to attainable goals.

Steward of Relationships

Relationships are a fundamental building block of leadership. As leaders, we are entrusted to build and maintain relationships with many stakeholders. First and foremost, we must build relationships with our students. Building professional, meaningful relationships with kids makes the entire educational process easier for all of us. Get to know your students, relate to them, and build them up.

We also must build relationships with our peers and faculty. We are all here for each other and our kids. Let's encourage one another and remember that our weaknesses are others' strengths. Learn from each other and grow.

The wise steward of relationships works with parents. Make contact and share successes so that when a failure happens, you have formed a relationship and can move forward together. Look to these questions as you steward leadership through relationships:

1. What are my weaknesses, and how can my relationship strengthen those weaknesses?
2. Can my strengths help those around me to be better?
3. How can this relationship make me a better leader?

An educational leader can utilize the stewardship of relationships in valuable ways. Leaders are most productive when building relationships. It is hard for anyone to be effective without actively pursuing growth in this area. We are not on an island all by ourselves, but we are here for each other and our students.

Burress suggests you may see a higher priority in one area over another and asks that, as the year begins, you reflect upon them and try to strengthen your stewardship as it relates to your students, staff, school, and community.

Burress's final comment is appropriate for teachers and students as it relates well to the chapter's definition of stewardship – Trust in yourself to be a wise steward, and remember, the wise steward leaves the things they are entrusted with in better shape than they were given.

Earth Day — Smart Practices

Every year on April 22, Earth Day is celebrated, marking the anniversary of the birth of the modern sentimental movement started in 1970.[5] In 1990, Earth Day became an International Day of Observance. Currently, over 192 countries observe Earth Day, and it is the largest secular observance in the world.[6]

Teaching students about Earth Day is vital to future generations' understanding of the importance of preserving the land on which we live. Each generation bears responsibility for what future generations will inherit. Education is the most powerful catalyst for change. Ideally, after learning about this day's importance, students will be inspired to act, whether that involves planting a tree, recycling at home, or picking up trash at a public park. We only get one Earth, so let's take care of it.[7]

The importance of stewardship has been presented to young learners as the vigilant care of something valuable. This builds on previous lessons and activities to focus children's responsibility on safe online behavior, caring for possessions, and their environment. This foundational approach aims to cultivate future custodians of the Earth by teaching respect and care for the planet.

The concepts of respect, care, and environmental stewardship play an important role in the child's online environment as well.

The concept of digital stewardship and a young learner's role in this stewardship is an essential educational lesson for students. Educating children on environmental and digital stewardship can link and reinforce the idea of being environmentally aware of the planet around them and cyber safety aware when in an online environment.

What Is Digital Stewardship?

Digital stewardship is the management of digital objects over the long term through careful digital asset management practices.[8] Digital stewardship also involves awareness about the digital records we create, how they are stored, and where we store them.[9]

Digital stewardship in the context of technology and cybersecurity refers to the responsible management and oversight of digital information and resources to ensure their safe, ethical, and efficient use. This concept encompasses a range of practices aimed at protecting the online environments in which we work, information (and by default data) integrity, and an individual's privacy. These practices are essential in creating a cyber-safe environment and are an important lesson for young learners embarking on initial online adventures and becoming good digital stewards.

Importance of Being Good Stewards in the Digital World

In discussing the role and importance of being a good steward in the digital world with young learners, address the concept simply and directly and work with analogies that resonate with children.

A good start is to remind children that being a good steward in the digital world means taking care of their online spaces and the resources available to them, just like they would take care of their toys and classrooms. It means being kind when they message or talk to others on the Internet, not sharing personal information like their home address or secret safe word, and asking a trusted person for help if they see something online that makes them feel confused or uncomfortable.

Ten Principles of Good Digital Stewardship

Discussing the following ten principles of good digital stewardship is an appropriate way to introduce the concept of digital stewardship and engage young learners in taking the first steps in becoming good digital stewards.

Principle #1: Respect Digital Devices

- Handle all digital devices with care.
- Use technology responsibly, and do not cause harm or damage to digital devices.
- Understand that through the improper or intentional misuse of technology (Internet, mobile phones, laptops, etc.), you could cause harm to people, even if you did not intend to do so.

Principle #2: Practice Responsible Online Behavior

- Adhere to the values of kindness, empathy, and respect in all digital interactions.
- Think before posting or sharing content (especially any personal information) online to ensure it is positive and appropriate and that you have permission to post it.

Principle #3: Protect Personal Information

- Remember the importance of privacy, especially the privacy of personal information – yours, your family's, and your friends.

- Never share personal information, such as your full name, home address, or school details, without first receiving permission from a parent or trusted adult.

Principle #4: Understand Online Consequences

- Realize that your every online action has consequences.
- Be an example by always practicing cyber-safe digital behavior, both personally and when with others online.

Principle #5: Identify Trusted Adults

- Identify and communicate with a trusted adult when you use the Internet.
- When you use the Internet, always tell a trusted adult if something makes you feel weird or not safe.
- Remember that it's okay to seek help or guidance from parents, teachers, or another trusted adult if you ever experience a problem when online and using the computer.

Principle #6: Develop Positive Digital Habits

- Work toward establishing positive online habits, such as balanced screen time, constructive content consumption, and using technology as a tool for learning and creativity.

Principle #7: Collaborate Safely Online

- Practice safe and positive collaboration in your online spaces.
- Remember that teamwork, respect for diverse opinions, and appropriate online communication are good characteristics everyone should practice and keep.

Principle #8: Understand Digital Footprints

- When you use the Internet, like playing games or learning, some information about you can be collected secretly, without you even knowing it.
- Before playing games online, make sure to set up and turn on your privacy settings.
- Remember, everything you do online is kept, so be careful about what you say, do, or share.
- Try to always be nice and make a good impression when you're on the Internet.

Principle #9: Become Critical Digital Consumers

- Develop critical thinking skills to evaluate online content.

Figure 4.1 Digital steward superhero.[10]

- Learn to differentiate between reliable and unreliable information (online or not).
- Always be cautious and aware of potential inherent technology risks (online and offline).

Principle #10: Strive for Continuous Cyber Safety Education

- Continue learning and practicing your cyber safety skills.
- Discuss with a trusted adult, the potential online dangers and strategies for staying cyber-safe.

By discussing and reviewing these principles with your students, you can help them become good stewards of the digital world, contributing to a positive, safe, and respectful online environment for themselves and others.

Explain to students that being a good digital steward and following the ten principles of good digital stewardship, is like being a superhero (see Figure 4.1). They protect themselves and their friends by making smart choices online and always being cyber-safe.

Being Good Digital Stewards — Leading by Example

Discussing the ten principles of good digital stewardship with young learners is the first step. Reinforcing these principles requires a thoughtful and age-appropriate approach. Creating memory cues that can help reinforce these principles is the next step.

Among the most effective methods to teach students the responsibility of good digital stewardship, reinforce learning and maximize retention is to use language that is easy for young learners to understand. Avoid technical jargon and explain concepts in simple terms. Incorporating age-appropriate stories or scenarios to illustrate digital stewardship concepts into classroom discussions is also very effective in reinforcing these important concepts.

The following are suggested examples of reinforcing the concepts of good digital stewardship embodied within the ten principles discussed previously.

Sharing Nicely Online:

- Example: When playing online games or using educational apps, share turns with friends and take turns using devices. This helps create a positive and inclusive digital environment.

Keeping Personal Information Private:

- Example: Explain to children that just as they don't share personal details with strangers in real life, they shouldn't share personal information online. Teach them to keep usernames, passwords, and other personal details private.

Being Kind in Online Messages:

- Example: Emphasize the importance of using kind words when chatting or messaging online. Encourage children to think about how their words might make others feel and to communicate in a friendly and respectful manner.

Asking for Help from Trusted Adults:

- Example: Teach children that if they see something online that makes them uncomfortable or if they are unsure about something, they should talk to a trusted adult, like a parent, teacher, or caregiver.

Taking Care of Devices:

- Example: Explain to children that just like they take care of their toys, they also need to take care of digital devices. This includes not dropping them, keeping them clean, and asking for help if something isn't working.

Using Screen Time Wisely:

- Example: Teach children about the importance of balanced screen time. Explain that while screens can be fun and educational, it's also essential to engage in other activities like playing outside, reading books, or spending time with family.

Being a Good Digital Friend:

- Example: Encourage children to be good digital friends by helping others, sharing positive content, and reporting anything that seems unkind or inappropriate. Help students understand the impact of their digital actions on others.

Respecting Others' Turns:

- Example: In educational settings where multiple children share devices, stress the importance of waiting patiently for their turn. This helps teach the concept of sharing resources and being considerate of others.

Creating Positive Digital Art:

- Example: If children engage in digital art or creative apps, encourage them to create positive and uplifting content. This fosters a sense of responsibility for the type of content they contribute to the online space.

Learning about Online Characters:

- Example: If children are interacting with digital characters or avatars, discuss the concept of treating virtual characters with the same kindness and respect they would show to real people.

These examples are not exhaustive; however, they will help children understand the basics of responsible digital stewardship in a way that is relatable and age-appropriate.

Developing a Stewardship Pledge

To foster awareness among students that this is a continual responsibility, teachers may work with students to develop a classroom stewardship pledge.

Discussions have centered around building character, making good choices, safe cyber habits, and the student's responsibility to act safely with possessions, with other people, and with the planet.

To create the classroom pledge, you may want to start by discussing respect and what it means. Respect takes many forms – to the student's belongings, to other people's belongings, to each other, to the community, and to the planet.

After discussing what respect means, a classroom pledge may be built on the word RESPECT.

Here's a classroom pledge that incorporates the concepts of character education, acceptable behavior, ethical behavior, practicing and maintaining safe cyber habits, and stewardship using the letters "R," "E," "S," "P," "E," "C," and "T."[11]

In the classroom, we pledge to:

R – "Reach out":

We promise to reach out to a teacher when the lights are left on in our classroom or water is left running, to do what is good for our environment.

E – "Engage":

We will engage in kind behavior with my classmates, both in person and online.

S – "Speak up":

We will speak up and let a teacher know when someone is being bullied or treated unkindly.

P – "Participate":

We commit to participating, in person and online, in kind and honest activities.

E – "Enjoy":

We will enjoy being respectful and being a model for our school and community.

C – "Conduct":

We will conduct ourselves in a way that shows we are kind and respectful to each other and to our environment.

T – "Teach":

We promise to teach other classrooms and our friends about our pledge to increase kindness and good character in our school, at home, and in our community.

As the students begin to comprehend and grasp the classroom stewardship pledge, a personal student stewardship pledge can be developed.

This pledge, developed at an early age, will lead to a sense of responsibility that will stay with the student into future years.

Here's a student pledge that incorporates the concepts of character education, acceptable behavior, ethical behavior, practicing and maintaining safe cyber habits, and stewardship using the letters "R," "E," "S," "P," "E," "C," and "T."[12]

R – "Respect":
"I pledge to show respect to everyone I meet, both online and offline."

E – "Ethical":
"I promise to be ethical and do what's right, even when no one is watching."

S – "Safe":
"I will keep myself safe online by only visiting websites my parents or teachers approve."

P – "Positive":
"I commit to spreading positive vibes and being a good friend to others."

E – "Environment":
"I will take care of our environment and be a good steward of the Earth."

C – "Character":
"I promise to build good character by being honest, kind, and responsible."

T – "Trustworthy":
"I will be trustworthy and make good choices in everything I do."

This pledge helps young children understand the importance of character education, acceptable and ethical behavior, practicing safe cyber habits, and stewardship while using simple words and concepts that are easy for kindergarten through second-grade students to comprehend and embrace.

Summary

This chapter presented the importance of teaching young children the concepts of digital stewardship and cyber safety. The chapter emphasized that just as children learn to be good stewards of the environment, they must also learn to be responsible custodians of their digital world.

Ten main principles of good digital stewardship were presented, covering such areas as handling devices carefully, thinking before posting, not sharing personal details, realizing that all actions have consequences, collaborating online safely, understanding digital footprints, and striving for continual cyber safety education.

To reinforce these principles, the authors suggest using clear, simple language and incorporating relatable stories and scenarios. Examples of good digital stewardship are provided on topics such as sharing nicely online, being kind in messages, asking adults for help, taking care of devices, balancing screen time, respecting turns, and treating virtual characters with respect.

Educators must lead by example, demonstrating responsible use of technology in an educational setting. Children are watching and are eager to mimic their teacher's behavior. Setting a good example helps children see positive digital stewardship in action.

Incorporating hands-on learning activities into daily class lessons and exercises greatly contributes to reinforcing the retention of core concepts.

By introducing principles of good digital stewardship in an age-appropriate way, reinforcing them through ongoing exercises, and encouraging open communication, young learners can become responsible, ethical caretakers of their online world. Mastering these concepts at an early age helps young learners build a strong foundation for ongoing cyber safety practices and exhibit positive digital citizenship behavior.

LESSON PLANS

Kindergarten

Exploring Cyber Safety through Stewardship

TOC Title: K Stewardship

Lesson Title: What Is Stewardship?

Grade Level: K

Duration: 60 minutes including read aloud. (Consider splitting this lesson plan into two sessions of 35 and 25 minutes each. A recommended break point is after the nature hunt.)

Objective:

- The students will understand the concept of stewardship and its importance in caring for the environment.
- Students will apply the concept of stewardship in a hands-on way.

Suggested Materials:

- Bags or baskets.
- Markers or crayons.
- Construction paper.
- Glue.
- Outdoor items collected from nature (if possible), e.g., leaves, pinecones, rocks, etc.
- Optional – pictures from nature settings that can be cut up for an art project if not using real objects.

Procedure

Introduction (10 minutes):

a. Ask students what it means to take care of something. Discuss examples like taking care of toys, pets, or plants.
b. Introduce the term "stewardship" and explain that it means taking care of something, and that today we will learn about stewardship of the environment.

Optional Read Aloud (10 minutes)

a. Suggested book: A Tree Is Nice by Janice May Udry

Activity – Nature Hunt (15 minutes):

a. Take students outside to look for and collect nature items found around school grounds (leaves, rocks, pinecones, etc.). Tell students to avoid picking things that are still growing. A pinecone on the ground is fine, but don't pull one off of a tree. Incorporate cleaning up litter if possible. If this is not an option, consider showing pictures of forests, prairies, deserts, and other natural environments so that students can identify naturally occurring objects that interest them.

This is a good breaking point if splitting into two sessions.

Art Mini-Project (20 minutes):

a. Students use the items they collected to create a nature collage.
b. If no items were collected, students could draw or cut out pictures to create their collage.

Conclusion (5 minutes):

a. Group reflection with suggested questions:
   - What did you learn about stewardship today?
   - How did it feel to help take care of nature during the nature hunt?
   - Why do you think it is important to take care of the Earth?

Assessment:

a. Observe students' participation during the nature hunt and their ability to connect the concept of stewardship with real-life actions.

Grade 1

Exploring Cyber Safety through Stewardship

TOC Title: Grade 1 Stewardship

Lesson Title: What Stewardship Means to Earth

Grade Level: 1

Duration: 40–50 minutes

Objective:

- The students will understand the concept of stewardship and its importance in taking care of the Earth.
- The students will be able to identify simple actions that will help them be good stewards of the environment.

Suggested Materials:

- Large world map or globe.
- Pictures or illustrations of Earth-friendly actions, e.g., recycling, saving water, and planting trees.
- Drawing paper and crayons/markers.
- Storybook on Earth stewardship. Suggested book: The Earth Book by Todd Parr or another book that addresses stewardship at an age-appropriate level.
- Optional materials if available: small pots, cups, soil, water, and seeds for planting activity (lettuce seeds are a fast-sprouting seed that work well).

Procedure

Introduction (10 minutes):

a. Begin by showing the world map or globe to the students. Ask them why Earth is important.
b. Introduce stewardship by explaining that it means taking care of something.
c. Ask for ideas on how we can take care of the Earth just like we take care of our toys or pets.

Read Aloud (10 minutes) [Optional if reading books are not available]

a. Read the suggested book or a similar one.

Class Discussion (5 minutes):

a. Show pictures of Earth-friendly activities. Discuss why each is important for taking care of the Earth.
b. Encourage students to share any actions they already take at home that take care of the Earth.

Activity (15 minutes):

a. Optional activity if materials are available:
b. Planting seeds.
c. Each student will fill a small pot or cup with soil. Bury a seed in the soil and add water. Enough to dampen. Keep pots/cups in a well-lit area and water them daily. Observe daily to watch for sprouts.

Alternate Activity (15 minutes):

a. Provide each student with drawing paper and crayons/markers.
b. Ask them to draw a picture of themselves doing something good for the Earth, such as recycling, picking up trash, or saving water.

Conclusion (5 minutes):

a. Summarize the importance of being good stewards of the Earth and how everyone can contribute to a healthier planet.
b. If students did the drawing activity, have them share.

Homework (Optional):

a. Encourage students to discuss the concept and importance of stewardship of the environment with their families.

Assessment:

a. Informal assessment can be done by observing student engagement in activities and discussions.

Grade 2

Exploring Cyber Safety through Stewardship

TOC Title: G 2 Stewardship

Lesson Title: How to Be a Good Steward

Grade Level: 2

Duration: 45 minutes

Objective:

- The students will understand stewardship and the various ways they can care for the environment.

Suggested Materials:

- Paper.
- Crayons, markers or colored pencils.
- Optional picture book. Suggested Book: Old Enough to Save the World by Loll Kirby or another book that addresses stewardship at an age-appropriate level.

Procedure

Introduction (10 minutes):

a. Begin with a class discussion on what it means to take care of something. Ask students if they have responsibilities at home, like cleaning their room, taking care of a pet, etc. Explain that stewardship is just like that – taking care of the Earth and all the living things on it.

Read Aloud (15 minutes):

a. Read the suggested or similar book.

Discussion:

a. Briefly discuss with the class what they learned about taking care of the Earth from the book.

Activity: Stewardship Poster (15 minutes):

a. Provide each student with a piece of paper and art supplies.

- Students draw and/or write about one or more things they can do to help take care of the environment.
- Encourage creativity. Students may use something they learned from the book or use their own ideas.
- Allow students to share their posters with the class afterward. Consider displaying student work around the classroom.

Conclusion (5 minutes):

a. Ask students to share what they have learned about stewardship, and why it is everyone's responsibility to take care of the Earth.

Homework (Optional):

a. Encourage students to share what they have learned with their families.

Assessment:

a. An informal assessment of student work and participation is suggested.

Notes

1 Tucker, D., Andrews, B., & Hayes, K., "Student-Driven Stewardship Projects Using Environmental Education to Promote the Use of Science Practices," National Science and Teaching Association, Student-Driven Stewardship Projects|NSTA, www.nsta.org/science-and-children/science-and-children-septemberoctober-2020/student-driven-stewardship-projects, retrieved December 15, 2023.
2 Lesley University, (n.d.), "Raising Children to be Good Stewards of the Earth,", retrieved December 14, 2023.
3 Fletcher, C., October 5, 2023, "The Importance of Environmental Education for a Sustainable Future," Earth.org, The Importance of Environmental Education for a Sustainable Future|Earth.Org, retrieved December 15, 2023.
4 Burress, D., May 15, 2020, "The Importance of Stewardship in Leadership," Graduate Programs for Educators, The Importance of Stewardship in Leadership - Graduate Programs for Educators, retrieved December 15, 2023.
5 https://www.earthday.org/, retrieved November 3, 2023.
6 Earth Org,, April 22, 2024, 5 Earth Day Facts to Know about and How to Get Involved|Earth.Org, retrieved May 18, 2024.
7 Educators4SC, (n.d.), "Teaching about Earth Day," https://educators4sc.org/teaching-about-earth-day/ retrieved December 19, 2023.
8 Harvard Library Digital Repository Service, (n.d.), "DRS Policy Guide, v3", Harvard University, p. 9, https://wiki.harvard.edu/confluence/download/attachments/215270190/drs_policy_guide_V_3_0.pdf?version=1&modificationDate=1492704676000&api=v2, retrieved January 8, 2024.
9 Oregon Heritage, July 2015, "Digital Stewardship & Curation," Oregon Heritage Bulletin #23, www.OregonHeritage.org, oregon.heritage@oprd.oregon.gov, retrieved January 8, 2024.
10 Marcella, A., 2024, Digital Steward Superhero, Image generated by DALL·E 2, prompt created by authors.
11 Marcella, A., 2024, "Create a Student Pledge Summarizing the Concepts of Character Education, Acceptable and Ethical Behavior, Practicing and Maintaining Safe Cyber Habits, and Stewardship," ChatGPT-4, developed by OpenAI, retrieved January 4, 2024.
12 Marcella, A., 2024, "Create a Student Pledge Summarizing the Concepts of Character Education, Acceptable and Ethical Behavior, Practicing and Maintaining Safe Cyber Habits, and Stewardship," ChatGPT-4, developed by OpenAI, retrieved January 4, 2024.

Appendix A: Teacher Resource List

The following materials represent a non-exhaustive list of supplemental teaching resources that the educator may find both of interest and use in developing class instructional materials to further the cyber-safe education of young learners. The resources are presented alphabetically, and the URL included with each resource is the most current as of February 2024.

Children's Illustrated Books

"Kidpower Children's Safety Comics Color Edition: Use Your Power to Stay Safe!" Have fun while teaching kids to be safe from bullying, abuse, and violence. The books have updated social stories and skills, including how to help kids stay safe online. [www.kidpower.org/books/kidpower-childrens-safety-comics-color-edition]

"Stand in My Shoes: Kids Learning about Empathy" by Bob Sornson – This book, while not a comic, uses engaging illustrations and storytelling to teach kids about empathy. [ISBN-13 – 978-0578807942]

"Kindness Is Cooler, Mrs. Ruler" by Margery Cuyler and Sachiko Yoshikawa is another illustrated book that is great for starting conversations about kindness, a key element of character education. [ISBN-13 –978-0689873447]

"Bullies Never Win" by Margery Cuyler and Arthur Howard – Standing up for oneself is the theme of this book. [ISBN-13 –978-0689861871]

Cyber Safety Research

Brito, R., & Dias, P., 2021, "Pedagogical Storytelling Material for Children Regarding Online Safety: Pilot Study in Kindergartens," IGI Global. https://doi.org/10.4018/978-1-7998-5770-9.ch011, www.igi-global.com/chapter/pedagogical-storytelling-material-for-children-regarding-online-safety/268220. [Highlights the importance of using storytelling as a pedagogical tool for online safety education for young children]

Finkelhor, D., Walsh, K., Jones, L., Mitchell, K., & Collier, A., 2021, "Youth Internet Safety Education: Aligning Programs with the Evidence Base," *Trauma, Violence, & Abuse*, 22(5), 1233–1247. https://doi.org/10.1177/1524838020916257. [This research reviews the literature on online harms to children and discusses how to develop effective online safety messages]

Hartikainen, H., Iivari, N., & Kinnula, M., December 2019, "Children's Design Recommendations for Online Safety Education," *International Journal of Child-Computer Interaction*, 22, 100146. ISSN 2212-8689, https://doi.org/10.1016/j.ijcci.2019.100146, www.sciencedirect.com/science/article/pii/S2212868917300764. [Focuses on involving children in the design of online safety education, ensuring that the content is relatable and understandable for them]

Moynihan, M., November 28, 2014, "Using Web 2.0 Tools to Teach Online Safety Education in the Intermediate Grades," Vancouver Island University, https://viuspace.viu.ca/bitstream/handle/10613/2328/MoynihanOLTD.pdf?sequence=3. [Demonstrates how Web 2.0 tools can be effectively used in teaching online safety to young students]

Ondrušková, D., & Pospíšil, R., 2023, "The Good Practices for Implementation of Cyber Security Education for School Children," *Contemporary Educational Technology*, 15(3), Article No. ep435. ISSN: 1309-517X (Online), https://doi.org/10.30935/cedtech/13253. [Outlines best practices for implementing cybersecurity education in schools, emphasizing age-appropriate methods]

Wishart, J., 2004, "Internet Safety in Emerging Educational Contexts," *Computers & Education*, 43, 193–204. https://doi.org/10.1016/j.compedu.2003.12.013, www.researchgate.net/publication/220140831_Internet_safety_in_emerging_educational_contexts/citation/download. [This paper reports on the consequent audit of Internet Safety practices in over 500 schools from 27 Local Education Authorities (LEAs) across England, commissioned by Becta and conducted during the summer term 2002]

Digital Resources

BrainPOP Jr. Internet Safety Video: A resource for teaching children about interacting with strangers online and protecting private information. Includes activities such as Word Play and Write about It, along with quizzes for assessment. [https://jr.brainpop.com/artsandtechnology/technology/internetsafety]

Georgia Public Broadcasting (GPB): GPB Education is Georgia's digital media content provider for the classroom, offering locally produced, Georgia-specific content and digital streaming services across all subject areas to teachers and students. GPB's goal is to remain at the forefront of the digital learning movement by creating, curating, and distributing quality educational programs and services through a state-of-the-art production facility, a cutting-edge digital media division, and a partnership with PBS LearningMedia. GPB Education also provides training for Georgia's educators and educational institutions. [www.gpb.org/education/learn/school-stories/science-in-action]

Mentoring Youth: Provides resources and tools for mentoring youth. [https://youth.gov/youth-topics/mentoring#:~:text=Mentoring%20provides%20youth%20with%20mentors,DuBois%20and%20Karcher%2C%202005)]

National Alliance on Mental Illness (NAMI): In addition to NAMI's support of mental illness, they provide resources for bullying, a teen and youth call center, stress management, and much more. [www.nami.org/Home]

Neuroscience in Education – Center for Educational Neuroscience: A Component of Mind, Brain and Education is an emerging scientific field that brings together researchers in cognitive neuroscience, developmental cognitive neuroscience, educational psychology, educational technology, education theory, and other related disciplines to explore the interactions between biological processes and education. [www.educationalneuroscience.org.uk/about-us/what-is-educational-neuroscience]

Instructional Resources

Common Sense Education – 23 Great Lesson Plans for Internet Safety: Provides lesson plans to help students build critical-thinking skills and navigate online dilemmas. [www.commonsense.org/education/articles/23-great-lesson-plans-for-internet-safety]

Common Sense Education Safety in My Online Neighborhood: Offers lesson plans for digital citizenship that address timely topics such as cyberbullying, cybersecurity, identity verification, physical security, and online safety. Their curriculum is research-backed and designed to help schools navigate these issues with their students. Lesson plans culminate in interactive activities like the Internet Traffic Light game, where students assess whether the presented online content is appropriate or not. [www.commonsense.org]

Edutopia: Provides a lesson plan for teaching Internet safety to students as young as kindergarten. The plan includes discussions about strangers online and protecting private information. Additional activities such as watching an Internet safety video, quizzes, and creating safety posters are also outlined. [www.edutopia.org]

EVERFI – The Compassion Project: A free empathy curriculum for 2nd–5th graders, designed to facilitate lessons around social and emotional skills. [https://everfi.com/k-12/character-education]

GoodCharacter.com – Character Education & Social-Emotional (SEL) Learning Resources: Provides teaching guides for K-12 on character education and social-emotional learning, covering topics like mindfulness, conflict resolution, and bullying.

- Bullying [www.livewiremedia.com/product-category/bullying]
- Citizenship Guides [www.livewiremedia.com/product-category/citizenship]
- Conflict Resolution [www.livewiremedia.com/product-category/conflict-resolution-2]

- Fairness [www.livewiremedia.com/product-category/fairness]
- Good Character Teaching Guides [www.livewiremedia.com/product-category/character-education] and [www.livewiremedia.com/product-category/character-counts]
- Manners and Politeness [www.livewiremedia.com/product-category/manners-politeness]
- Mindfulness [www.livewiremedia.com/product-category/mindfulness]
- Respect [www.livewiremedia.com/product-category/respect]
- Social Media & Internet Safety [www.livewiremedia.com/product-category/social-media-internet-safety-2]

Learning.com: Provides a lesson titled "Online Safety: Netiquette and Cyberbullying for Grades K-2," which teaches students about netiquette, proper online behavior, and cyberbullying prevention. Teachers can download this lesson plan to use in their classrooms. [equip.learning.com]

MiTechKids Internet Safety Lesson: This lesson for 2nd graders focuses on staying safe online, not talking to strangers, and reporting inappropriate content. It includes a Brain Pop video and various activities. [www.remc.org/mitechkids]

Safer, Smarter Kids Curriculum: A comprehensive program that covers various safety topics, including cyber safety, body boundaries, and recognizing unsafe situations. [https://safersmarterkids.org]

TeacherVision Safety Printables and Activities: Offers a wide array of lesson plans, printables, and resources, including content for teaching Internet safety. [www.teachervision.com/subjects/health-safety/safety]

UNODC Primary Lesson Plan for Staying Safe Online: While this resource is more general, it does include exercises and additional teaching tools that can be adapted for younger students. [www.unodc.org/e4j/en/primary/e4j-tools-and-materials/lesson-plan_staying-safe-online.html]

Interactive Assessment Exercises

Cyber Awareness & Safety Education (CASE): This curriculum offers a series of interactive classes where students actively learn through videos, case studies, and group discussions. The end-of-course test measures mastery of the material. This program includes pretests and post-tests to assess student understanding of Internet safety, along with active learning techniques to personalize the topics covered. [www.cybersafetyconsulting.com]

Interactive Games and Classroom Exercises

CISA Cybersecurity Awareness Program Student Resources: Offers downloadable activity sheets, books, and other resources for young children on cyber safety. It also includes games from the FBI's Safe Online Surfing program and tips from the National Cyber Security Alliance.

- Savvy Cyber Kids [https://savvycyberkids.org]
- Safe Online Surfing by the FBI [https://sos.fbi.gov/en]
- NetSmartz Kids [www.netsmartzkids.org]

Digital Well-Being Packet and Classroom Activities from Be Internet Awesome: Is a multifaceted program that includes both a game called Interland to engage students in learning about online safety and a curriculum for educators to teach kids how to be safe and responsible online. They also offer a pledge for families to promote online safety at home.

The site provides comprehensive guides and printable activities to support educators in introducing digital well-being to students. Lesson plans for five topics with activities and worksheets designed to complement the Interland game, focusing on being smart, alert, strong, kind, and brave online. Provides interactive, game-based activities that assess students' understanding of Internet safety in a fun and engaging way. [https://beinternetawesome.withgoogle.com/en_us/educators]

Equip Learning – Online Safety: Netiquette and Cyberbullying for Grades K-2: A lesson plan that introduces students to netiquette, proper online behavior, and cyberbullying prevention.

The plan includes creating a poster that aligns with educational standards. English Language Arts and ISTE standards. The activities can also serve as an assessment of the student's understanding of online behavior and safety. [https://equip.learning.com]

Safety Land by AT&T: An interactive game for elementary students that teaches Internet safety in a fun and engaging way. [https://freetech4teach.teachermade.com/2011/04/at-safety-land-cyber-safety-game-for]

TED-Ed Video Resources

Note: TED-Ed Is TED Talk's [www.ted.com/talks] youth and education initiative.

"3 Questions to Ask Yourself about Everything You Do" by Stacey Abrams: "How you respond to setbacks is what defines your character," a talk that could be adapted to discuss character and self-reflection, important components of character education. [https://ed.ted.com/search?qs=%223+questions+to+ask+yourself+about+everything+you+do%22+]

"Grit: The Power of Passion and Perseverance" by Angela Lee Duckworth: This talk may be adapted to be used to discuss the character trait of perseverance. [https://ed.ted.com/search?qs=Grit%3A+The+Power+of+passion+and+perseverance]

"How to Build Your Confidence – And Spark It in Others" by Brittany Packnett: This talk can help in discussions with students about self-confidence as a part of one's character. [https://www.youtube.com/watch?v=b5ZESpOAolU&t=4s]

YouTube Video Resources

"Being Respectful Video for Kids"|Character Education – by Jessica Diaz: This video can help teach kids about being respectful, an essential aspect of character education. [https://www.youtube.com/watch?v=KxnxObAyfSA]

"Consequences for Kids"|Character Education – by Jessica Diaz: This video is designed to help children understand what consequences are and why they matter, reinforcing the concept of accountability. [www.youtube.com/watch?v=LLZZYf_mlOA]

"Cyber Security for Kids" – by Neel Nation: This video is about being safe on the Internet. The video provides five tips for kids to be safe on the Internet. Being cyber-safe is very important, as kids are always online these days. This video talks about the importance of passwords, protecting personal information, limiting screen time, etc. [www.youtube.com/watch?v=nVEyG3C-Mqw]

"Cyberbullying – How to Avoid Cyber Abuse" – by Smile and Learn: This video showcases a series of situations related to how students can stop cyberbullying. Also discussed is the concept of being a good friend, not a bystander, and respecting privacy, and how actions can make a difference. [www.youtube.com/watch?v=dMdKmHjpgFk]

"Cybersecurity Training for Kids" – by Malware Bytes: A fun video just for the kids, where they can learn all about Dr. Evil and her Internet schemes. [www.youtube.com/watch?v=XiU72Vzs5Is]

"Internet Safety for Kids K-3" – by Indiana University of Pennsylvania: This video is aimed at younger children and provides guidance on staying safe on the Internet. [www.youtube.com/watch?v=89eCHtFs0XM]

"Internet Safety Tips for Kids" – by Money Moments (MidFirst Bank): This short video covers safety tips for your kids using the Internet. [www.youtube.com/watch?v=qtJNRxMRuPE]

"Online Privacy for Kids - Internet Safety and Security for Kids" – by Smile and Learn: This video discusses the importance of online privacy and safety for children, which is crucial for their well-being in the digital age. [www.youtube.com/watch?v=yiKeLOKc1tw]

"Online Safety Staying Safe Online" – by Born in the USA: This resource is featured in Discovery Education Espresso's Online safety module. Learn how to stay safe and act responsibly when using the Internet in school and at home. [www.youtube.com/watch?v=PtfEnh0gbbU]

"Private and Personal Information" – by Common Sense Education: It is natural for students to enjoy sharing and connecting with others. However, sharing information online can sometimes come with risks. This video addresses these risks. [www.youtube.com/watch?v=MjPpG2e71Ec]

"Responsible Use of Technology for Kids" – by Smile and Learn: This video discusses how to make responsible use of technology, the Internet, and social media. In this compilation of videos about the responsible use of technology, the little ones will discover how to use their first phone responsibly, how to avoid cyberbullying, how to know what fake news is and how to spot it, and also how to protect their online privacy. [www.youtube.com/watch?v=JkkTN0pQ_Ug]

"What Is NETIQUETTE?" Internet Behavior Rules for Kids, Episode 1 – by Smile and Learn: This presentation discusses what netiquette is and how it can help children behave and interact correctly on the Internet. In this first series of two videos, kids will learn a basic set of rules to be polite and respectful on social networks and on the Internet in general. [www.youtube.com/watch?v=kZOfLN4YqhY]

"What is NETIQUETTE?" Internet Behavior Rules for Kids, Episode 2 – by Smile and Learn: In this second video of the series, kids will learn how to interact with other people on the Internet and see some tips on how not to share personal information or trust strangers. [www.youtube.com/watch?v=zhIm-CDJBpc]

Appendix B: Recommended Readings

The following titles represent a non-exhaustive list of books that the educator may find both of interest and use in furthering the cyber-safe education of young learners. Titles are presented alphabetically, and the ISBN 13 for each title is the most current as of February 2024.

"But It's Just a Game" by Julia Cook: This book is a great conversation starter about screen time and Internet safety, as it follows the story of a young boy who learns to balance his life online and offline. [ISBN 13: 978-1937870164]

"Chicken Clicking" by Jeanne Willis and Tony Ross: This story is about a young chick who discovers the joys and dangers of the Internet, providing a gentle introduction to online safety for young children. [ISBN 13: 978-1783441617]

"Click, Clack, Moo: Cows That Type" by Doreen Cronin and Betsy Lewin: Although not specifically about online safety, this book is a great introduction to technology and communication, themes that are foundational to understanding online behavior. [ISBN 13: 978-1481465403]

"Developing Mentoring and Coaching Relationships in Early Care and Education: A Reflective Approach (Practical Resources in ECE)" by Marilyn Chu: Is the ideal resource for anyone charged with guiding teachers as they encounter real-world challenges in today's early childhood programs. [ISBN 13: 978-0132658232]

"Do Unto Otters: A Book about Manners" by Laurie Keller: A playful take on the Golden Rule, teaching children about respect and kindness. [ISBN 13: 978-0312581404]

"Dot" by Randi Zuckerberg: A fun and colorful story about a tech-savvy girl who learns to unplug and play outside, highlighting the importance of a healthy balance between technology use and real-world experiences. [ISBN 13: 978-0062287519]

"Goldilocks and the Three Bears: A Hashtag Cautionary Tale" by Jeanne Willis and Tony Ross: A modern twist on the classic fairy tale, this book cleverly integrates online safety messages into the story. [ISBN 13: 978-1783448784]

"Goodnight iPad: A Parody for the Next Generation" by Ann Droyd: A playful take on the classic "Goodnight Moon," this book humorously addresses the modern challenges of too much screen time. [ISBN 13: 978-0399158568]

"Have You Filled a Bucket Today?" by Carol McCloud: This picture book encourages positive behavior as children see how rewarding it is to express daily kindness, appreciation, and love. [ISBN 13: 978-0996099936]

"It's OK to be Different: A Children's Picture Book about Diversity and Kindness" by Sharon Purtill and Sujata Saha: As more children read this book and learn this concept, we can positively impact the world while at the same time teaching early literacy. This non-fiction book closes with a short survey for children about kindness and is a terrific way to help you start an age-appropriate conversation about diversity. [ISBN 13: 978-0973410457]

"Make Your Brain Work" by Amy Brann: While this book is not written for K-2 students, it is a helpful tool for teachers and leaders to understand the latest insights from neuroscience about how our mind works. [ISBN 13: 978-1789660494]

"Nerdy Birdy Tweets" by Aaron Reynolds and Matt Davies: This book is a humorous take on social media etiquette, focusing on the importance of kindness and thinking before posting. [ISBN 13: 978-1626721289]

"Nettie and Webby: Take Care How You Share" by Wendy Goucher: Nettie and her cousin Claudia explore Cyberland, befriend Puffles, and learn the importance of taking care of how we share! [ISBN 13: 978-1803694962]

"Nettie in Cyberland: Introduce Cyber Security to Your Children (The Little Helpers)" by Wendy Goucher: This book takes Nettie and their friend Webby on her first adventure in Cyberland, where they find that not everything is cute and fluffy. Using the story of Nettie and Webby, cybersecurity is introduced to young children and starts the conversation about going online safely between the child and their adult reader. [ISBN 13: 978-1800319844].

"Norton and Alpha" by Kristyna Litten: A charming tale about technology, innovation, and ethics, great for sparking conversations about creativity and digital responsibility. [ISBN 13: 978-1454924999]

"Once Upon a Digital Story" by Amanda Hovious: This book is designed to help children learn to tell their own stories safely and creatively online. [ISBN 13: 9781516553341]

"Once Upon a Time… Online" by David Bedford and Rosie Reeve: A modern fairy tale where classic storybook characters learn about the Internet, perfect for discussing the dos and don'ts of online behavior. [ISBN 13: 978-1472392350]

"Online Safety for Children and Teens on the Autism Spectrum: A Parent's and Care's Guide" by Nicola Lonie: While this is more of a guide for parents and caregivers, it offers valuable insights and strategies that can be translated into simpler concepts for young children. [ISBN 13: 978-1849054425]

"Penguinpig" by Stuart Spendlow and Amy Bradley: This is a cautionary tale about a curious piglet who learns about the importance of staying safe online and not believing everything you read on the Internet. [ISBN 13: 978-0955926242]

"Screen Time Is Not Forever" by Elizabeth Verdick and Marieka Heinlen: This is a simple book that helps young readers understand the need for limits on screen time. [ISBN 13: 978-1631985379]

"Security Awareness Design in the New Normal Age" by Wendy Goucher: People working in our cyber world have access to a wide range of information, including sensitive personal or corporate information, which increases the risk. This book will primarily consider how knowledge about secure practice is not only understood and remembered but also reliably put into practice, even when a person is working alone. [Kindle ASIN – B0B1W1Z9LB]

"Shh! We Have a Plan" by Chris Haughton: This book, while not directly about online safety, teaches about the importance of planning and thinking ahead, a skill that's crucial when navigating the online world. [ISBN 13: 978-0763672935]

"The Berenstain Bears' Computer Trouble" by Jan and Mike Berenstain: A story from the beloved Berenstain Bears series that addresses issues like too much screen time and the need for balance. [ISBN 13: 978-0060573942]

"The Digital Citizenship Handbook for School Leaders: Fostering Positive Interactions Online" by Mike Ribble and Marty Park: Aimed at educators, this book contains valuable resources that can be adapted into lesson plans for young students. [ISBN 13: 978-1564847829]

"The Fabulous Friend Machine" by Nick Bland: This is a story about Popcorn the chicken, who discovers the Internet and learns valuable lessons about strangers online. [ISBN 13: 978-9352755356]

"The Technology Tail: A Digital Footprint Story" by Julia Cook: This engaging book teaches children about the lasting impact of their online behavior and the importance of managing their digital footprints. [ISBN 13: 978-1944882136]

"Webster's Email" by Hannah Whaley: This charming book tells the story of Webster and his first email, teaching children the importance of thinking before sending messages online. [ISBN 13: 978-0993001208]

Appendix C: Ideas for Certificates or Badges

The following is a non-exhaustive list of popular platforms for creating digital badges. The sites are presented alphabetically, and the URL included with each site is the most current as of February 2024.

- Google Slides: https://docs.google.com/presentation
- Open Badges by Mozilla: https://openbadges.org
- Credly: www.credly.com
- Badgr: https://badgr.com
- Accredible: www.accredible.com
- Canva: https://www.canva.com
- Adobe Spark: https://spark.adobe.com

These platforms can help you create, issue, and manage digital badges or certificates to recognize and reward students for demonstrating good character skills, acceptable behavior, and safe cyber habits.

Educators may wish to adapt the following certificate and badge ideas to suit the specific needs and preferences of your kindergarten to second-grade students.

These badges can serve as incentives and reminders for promoting positive character traits, ethical behavior, and safe cyber habits in young learners.

Character Education Certificates/Badge Ideas

- Caring Comrade
- Compassionate Classmate
- Cooperative Kid
- Courageous Character
- Empathy Champion
- Fairness Enthusiast
- Generosity Guru
- Good Citizenship Award
- Helpful Heart Award
- Honest Hero
- Integrity Star
- Kindness Ambassador
- Mindful Mini
- Patience Paladin
- Perseverance Pro
- Positive Attitude Expert
- Respectful Role Model
- Responsible Friend
- Self-Control Specialist
- Teamwork Titan

Acceptable and Ethical Behavior Certificates/Badge Ideas

- Acceptable Behavior Achiever
- Cyber Citizenship Certificate
- Cyberbullying Prevention Champion
- Digital Decision-Maker
- Digital Etiquette Pro
- Ethical Emoji Expert
- Honest Online Explorer
- Internet Safety Superstar
- Netiquette Ninja
- Online Friendship Advocate
- Privacy Protector
- Respectful Tech User
- Responsible Screen Time Star

- Safe Online Sharer
- Screen Time Role Model
- Social Media Manners Master
- Thoughtful Texting Trophy
- Trustworthy Technology User
- Virtual Kindness Crusader
- Wise Web Wanderer

Safe Cyber Habits Certificates/Badge Ideas

- App Safety Advocate
- Cyber Hygiene Hero
- Cyber Safety Scout
- Cyberbullying Defender
- Data Protection Dynamo
- Digital Detective
- Digital Privacy Pro
- Internet Explorer Expert
- Malware Awareness Master
- Online Safety Steward
- Online Security Sentinel
- Password Protector
- Phishing Prevention Pioneer
- Safe Clicker Award
- Safe Email Sender
- Safe Online Gamer
- Safe Surfing Star
- Secure Selfie Star
- Social Media Security Star
- Strong Password Superhero

Glossary of Terms & Definitions

Accessibility

Refers to the design of apps, devices, materials, and environments that support and enable access to content and educational activities for all learners. In addition to enabling students with disabilities to use content and participate in activities, the concepts also apply to accommodate the individual learning needs of students, such as English language learners, students in rural communities, or students from economically disadvantaged homes. Technology can support accessibility through embedded assistance, for example, text-to-speech, audio and digital text formats of instructional materials, programs that differentiate instruction, adaptive testing, built-in accommodations, and assistive technology.

Artificial Intelligence

A system that exhibits reasoning and performs some sort of automated decision-making without the interference of a human.

Authentic Learning Experiences

Experiences that place learners in the context of real-world experiences and challenges.

Behavior

A person's external reaction to their environment. All behaviors, such as yelling, crying, running, or throwing something, can be observed and measured.

Belittlement

The use of online forums, apps, or doctored images to try to insult and spread fake rumors, gossip, or untruths about a person.

Bioregion

An area whose limits are naturally defined by topographic and biological features (such as mountain ranges and ecosystems) and not political or religious boundaries.

Blended Learning

In a blended learning environment, learning occurs online and in person, augmenting and supporting teacher practice. Blended learning often allows students to have some control over the time, place, path, or pace of learning. In many blended learning models, students spend some of their face-to-face time with the teacher in a large group, some face-to-face time with a teacher or tutor in a small group, and some time learning with and from peers. Blended learning often benefits from a reconfiguration of the physical learning space to facilitate learning activities, providing a variety of technology-enabled learning zones optimized for collaboration, informal learning, and individual-focused study.

Cohesiveness

The degree to which group members enjoy collaborating with the other members of the group and are motivated to stay in the group.

Conformity

Adjusting one's behavior to align with the norms of a particular group.

Counterfactual

Contrary to facts or thinking about alternative possibilities for past or future events: what might happen/ have happened if…?

Critical Thinking

The act or practice of thinking by applying reason and questioning assumptions to solve problems and evaluate information without the influence of biases, personal feelings, and opinions.

Crosscutting Concepts

Ideas that hold true across many and often disparate disciplines.

Cyberbullying

Bullying is unwanted, aggressive behavior that involves a real or perceived power imbalance. This power imbalance can be physical. It can also revolve around popularity or the bully having access to embarrassing information about the victim. Generally, bullying is a repeated behavior or has the potential to be repeated. Cyberbullying, then, is when these bullying behaviors occur online, either through messaging, social media, or other digital channels.[1]

Cyberstalking

This highly intimidating form of cyberbullying is when a person tracks another in the digital sphere and sends them negative comments, which can include threats, to frighten and terrorize them. In some cases, this can even lead to physical stalking in real life.[2]

Data Point

A discrete unit of information.

Dataset

Related data points in a collection, usually with tags (labels) and a uniform order.

Digital Citizen

Refers to an individual who engages responsibly, ethically, and positively in the digital world. Being a digital citizen involves using technology, especially the Internet, in a manner that respects the rights and well-being of others while also contributing to the overall betterment of the online community. Digital citizenship encompasses a range of skills, knowledge, and attitudes that empower individuals to navigate the digital landscape safely and effectively. This includes understanding and practicing concepts such as online etiquette, responsible use of information, digital literacy, and awareness of potential risks and challenges in the digital environment.

Digital Citizenship

The safe, ethical, responsible, and informed use of technology. This concept encompasses a range of skills and literacies that can include Internet safety, privacy and security, cyberbullying, online reputation management, communication skills, information literacy, and creative credit and copyright.

Digital Citizenship (2)

The ability to use digital technology and media in safe, responsible, and ethical ways. Digital citizenship is a set of fundamental digital life skills that everyone needs to have.

Digital Competitiveness

The ability to solve global challenges, innovate, and create new opportunities in the digital economy by driving entrepreneurship, jobs, growth, and impact.

Digital Creativity

The ability to become a part of the digital ecosystem, and to create new knowledge, technologies, and content to turn ideas into reality.

Digital Intelligence (DQ)

A comprehensive set of technical, cognitive, meta-cognitive, and socio-emotional competencies grounded in universal moral values that enable individuals to face the challenges of digital life and adapt to its demands. Thus, individuals equipped with DQ become wise, competent, and future-ready digital citizens who successfully use, control and create technology to enhance humanity.[3]

Digital Use Divide

Traditionally, the digital divide referred to the gap between students who had access to the Internet and devices at school and home and those who did not. Significant progress is being made to increase Internet access in schools, libraries, and homes across the country. However, a digital use divide separates many students who use technology in ways that transform their learning from those who use the tools to complete the same activities but now with an electronic device (e.g., digital worksheets and online multiple-choice tests). The digital use divide is present in both formal and informal learning settings and across high- and low-poverty schools and communities.[4]

Digital Natives

Young people who have been born into a virtual reality, view the world differently, and have a 'digital footprint,' process infographics speedily, but lack basic capacity for interpersonal interactions.

Dissing

When someone spreads false or negative information about a person to damage their reputation.

Doxxing

See Trickery

Equity

In education means increasing all students' access to educational opportunities with a focus on closing achievement gaps and removing barriers that students face based on their race, ethnicity, or national origin; sex; sexual orientation or gender identity or expression; disability; English language ability; religion; socioeconomic status; or geographical location.

Flaming

When extreme or offensive language is used to try and cause stress or anxiety to the victim.

Fraping

When someone pretends to be someone else online and posts something silly or inappropriate. It can be intended as a joke, but in some cases, it can cause serious negative repercussions.

Generative Artificial Intelligence

Technology that creates content – including text, images, video, and computer code – by identifying patterns in large quantities of training data and then creating new, original material that has similar characteristics. Examples include ChatGPT for text and DALL-E and Midjourney for images.

Harassment

Constant online abuse which can occur on messaging apps or via comments on social media sites, chat rooms, or gaming sites.

Identity Theft

When a cybercriminal steals someone's personal information and uses it to assume their identity. This can involve the criminal applying for credit and loans, or even filing taxes using the victim's identity, potentially damaging their credit status.

Impersonation

Creating a fake social media account or email address to pretend to be someone else and then using it for negative purposes.

Machine Learning (ML)

A subcategory of artificial intelligence, in which the study or the application of computer algorithms is designed to improve automatically through experience. Machine learning algorithms build a model based on training data in order to perform a specific task, like aiding in prediction or decision-making processes, without necessarily being explicitly programmed to do so.

Masquerading

See Impersonation

Mentor

A role model for the child. Offers encouragement and support, and the child trusts the mentor to assist them with difficult situations or concepts.

Multi-Tiered System of Supports (MTSS)

A proactive and preventative framework that integrates data and instruction to maximize student achievement and support students' social, emotional, and behavioral needs from a strengths-based perspective.[5]

Norms

The acceptable standards of behavior within a group are shared by the members.

Neural Network

Also known as an artificial neural network, neural net, or deep neural net; a computer system inspired by living brains.

Neuroscience

Examines the structure and function of the human brain and nervous system. The study of neuroscience in education and learning is rapidly developing.

Openly Licensed Educational Resources

Teaching, learning, and research resources that reside in the public domain or have been released under a license that permits their use, modification, and sharing with others. Open resources may be full online courses, digital textbooks, or more granular resources such as images, videos, and assessment items.

Outcomes

Something that follows as a result or consequence, a conclusion reached through a process of logical thinking.

Outing

See Trickery.

Password

A confidential and alphanumeric sequence of characters that serves as a form of authentication, allowing an individual to access a computer system, online account, or digital device. The user must provide the correct password, which is known only to them, to verify their identity and gain authorized entry. Passwords play a crucial role in securing digital information and preventing unauthorized access to personal or sensitive data.

Password Hygiene

Creating unique and separate passwords for sensitive online accounts, managing passwords using browser or stand-alone applications, and the tactics of changing passwords.

Passphrase

A sequence of words or other text used as a form of authentication to access a computer system, online account, or digital device. Unlike traditional passwords, passphrases are typically longer and consist of multiple words or a combination of words, numbers, and symbols. Passphrases enhance security by creating a more complex authentication mechanism, making it harder for unauthorized individuals or automated programs to guess or crack the access credentials.

Personalized Learning

Refers to instruction in which the pace of learning and the instructional approach are optimized for the needs of each learner. Learning objectives, instructional approaches, and instructional content (and its sequencing) may all vary based on learner needs. In addition, learning activities are meaningful and relevant to learners, driven by their interests, and often self-initiated.

Pervasive Computing

Also known as ubiquitous computing, it refers to the integration of computing technology into everyday objects and environments, making them smart, connected, and capable of interacting with each other and with users seamlessly. The goal of pervasive computing is to create an environment where computational power is embedded in the fabric of daily life, making it unobtrusive and enhancing user experiences.

Pervasive Education

Typically, it refers to the integration and widespread presence of educational technologies, digital tools, and learning opportunities throughout various aspects of a learner's life and environment. It encompasses the idea that education is not confined to traditional classroom settings but extends into the broader context of a learner's daily experiences.

Pervasive Technology

Refers to the widespread integration and seamless presence of technology in various aspects of daily life and across diverse environments. In a pervasive technology landscape, digital tools, devices, and systems become an integral and often unnoticed part of the fabric of society, enhancing and influencing the way people live, work, and interact.

Positive Behavioral Interventions and Supports (PBIS)

A framework for creating safe, positive, equitable schools where every student can feel valued, connected to the school community, and supported by caring adults.[6]

Project-Based Learning

Where learning takes place in the context of authentic problems, continues over time, and brings in knowledge from many subjects. Project-based learning, if properly implemented and supported, helps students develop 21st-century skills, including creativity, collaboration, and leadership, and engages them in complex, real-world challenges that help them meet expectations for critical thinking.

Response to Intervention (RTI)

A data-driven approach schools use to support students academically. This approach is characterized by analyzing assessment data, planning high-quality instruction, implementing interventions, and evaluating each student's response to interventions.[7]

Role

A set of expected behavior patterns attributed to someone occupying a given position in a social unit.

Social and Emotional Learning (SEL)

A strengths-based developmental process that begins at birth and evolves across the lifespan. The process through which children, adolescents, and adults learn skills to support healthy development and relationships.

Trickery

When someone is groomed to build a relationship of trust and trick them into providing personal details or photos, which are then used to humiliate the person.

Trolling

Making constant comments to try and get a reaction out of the person. Some trolling can be intended as a harmless joke, but often it has a malicious intent.

Trustworthy Artificial Intelligence (AI)

AI systems that exhibit characteristics like resilience, integrity, security, and privacy if they're going to be useful and people can adopt them without fear.[8]

References

The terms and definitions identified in this glossary are based on a series of international reference documents and were developed, in part, using selected material from but not limited to, the following resources.

1. Karppinen, I., Nurse, J. R. C., & Varughese, J., 2023, "Oh Behave! The Annual Cybersecurity Attitudes and Behaviors Report 2023," The National Cybersecurity Alliance and CybSafe, www.cybsafe.com/whitepapers/cybersecurity-attitudes-and-behaviors-report.
2. Virgin Media, (n.d.), "How to Introduce the Kids to the Internet," www.virginmedia.com/blog/online-safety/childrens-internet-safety-test.
3. Park, Y., 2019, DQ Global Standards Report 2019, Common Framework for Digital Literacy, Skills and Readiness, DQ Institute, DQGlobalStandardsReport2019.pdf (dqinstitute.org).
4. U. S. Department of Education, January 2017, "Reimagining the Role of Technology in Education: 2017 National Education Technology Plan Update," U.S. Department of Education, Office of Educational Technology, Washington, D.C., Department of Education, http://tech.ed.gov.
5. American Institutes for Research, 2023, "Essential Components of MTSS," https://mtss4success.org/essential-components#:~:text=A%20multi%2Dtiered%20system%20of,from%20a%20strengths%2Dbased%20perspective.
6. Center on Positive Behavioral Interventions & Supports, 2023, "What Is PBIS?" www.pbis.org., www.pbis.org/pbis/what-is-pbis.
7. Enriching Students, 2021, "RTI vs. MTSS," Interval Technology Partners, LLC, www.enrichingstudents.com/rti-vs-mtss/#:~:text=RTI%20is%20considered%20a%20more,%2C%20and%20social-emotional%20support.
8. NIST, January 21, 2020, "Trustworthy AI: A Q&A with NIST's Chuck Romine," www.nist.gov/blogs/taking-measure/trustworthy-ai-qa-nists-chuck-romine.

Index

Printed in the United States
by Baker & Taylor Publisher Services